Critical Environments

Edited by

Sandra Buckley

Michael Hardt

Brian Massumi

THEORY OUT OF BOUNDS

...UNCONTAINED

BY

THE

DISCIPLINES,

INSUBORDINATE

PRACTICES OF RESISTANCE

...Inventing,

excessively,

in the between...

PROCESSES

OF

HYBRIDIZATION

Critical Environments

Postmodern Theory and the Pragmatics of the "Outside"

Cary Wolfe

Theory out of Bounds *Volume 13*

University of Minnesota Press

Minneapolis • London

Part of the Introduction originally appeared in different form in the essay "Introduction: The Politics of Systems and Environments," *Cultural Critique*, volume 30 (spring 1995). Used by permission of Oxford University Press. The material on Walter Benn Michaels in chapter 1 originally appeared in somewhat different form in *Discovering Difference: Contemporary Essays in American Culture*, ed. Christoph K. Lohmann (Bloomington: Indiana University Press, 1993). Some of the material on Richard Rorty in chapter 1 originally appeared in somewhat different form in the essay "Making Contingency Safe for Liberalism: The Pragmatics of Epistemology in Rorty and Luhmann," *New German Critique*, volume 61 (winter 1994). © 1994 Telos Press, Ltd. Permission to reprint has kindly been granted by Telos Press, Ltd. The material on Stanley Cavell in chapter 1 previously appeared in somewhat different form in the essay "Alone with America: Cavell, Emerson, and the Politics of Individualism," *New Literary History*, volume 25, number 1 (winter 1994). Permission to reprint has kindly been granted by Johns Hopkins University Press. The material on Niklas Luhmann in chapter 2 originally appeared in somewhat different form in the essay already cited from *New German Critique*, volume 61 (winter 1994). © Telos Press, Ltd. Permission to reprint has kindly been granted by Telos Press, Ltd. The remainder of chapter 2 originally appeared in somewhat different form in the essay "In Search of Post-Humanist Theory: The Second-Order Cybernetics of Maturana and Varela," *Cultural Critique*, volume 30 (spring 1995). Used by permission of Oxford University Press.

Published by the University of Minnesota Press
111 Third Avenue South, Suite 290
Minneapolis, MN 55401-2520

http://www.upress.umn.edu

Printed in the United States of America on acid-free paper

LIBRARY OF CONGRESS CATALOGING-IN-PUBLICATION DATA
Wolfe, Cary.
Critical environments : postmodern theory and the pragmatics of
the outside / Cary Wolfe.
p.   cm. — (Theory out of bounds ; v. 13)
Includes index.
ISBN 0-8166-3018-6 (hc : alk. paper). — ISBN 0-8166-3019-4 (pbk. :
alk. paper)
1. Postmodernism.   2. Pragmatism.   3. Systems theory.
4. Poststructuralism.   5. Theory (Philosophy)   6. Philosophy,
Modern—20th century.   I. Title.   II. Series.
B831.2.W64   1998
149'.97—dc21   97-34675

10   09   08   07   06   05   04   03   02   01   00   99   98   10   9   8   7   6   5   4   3   2   1

For Luccia

Contents

Acknowledgments

Exchanges with countless friends, colleagues, and students over the past several years have helped me worry, focus, and refine the set of problems and texts that form the topos of this study. I'd like to thank for their intelligence and generosity the members of my graduate seminar in the spring of 1995 at Indiana University on pragmatism and postmodernism, the members of SLAG (Science and Literature Affinity Group) at Indiana with whom I had a chance to discuss some of these texts, and my colleagues Richard Nash, Stephen Kellert, Jonathan Elmer, Lee Sterrenberg, Marti Crouch, Eva Cherniavsky, Jim Naremore, Oscar Kenshur, and Eva Knodt, some of whom provided valuable comments on earlier versions of the chapters.

I'd also like to thank Michael Bérubé and John McGowan for very detailed and helpful comments on an earlier draft, Niklas Luhmann, Katherine Hayles, and Marjorie Levinson for singularly interesting conversations along the way, and Donna Przybylowicz and Abdul JanMohamed for supporting publication of a related project in the pages of *Cultural Critique*.

Support of the more specifically monetary kind was provided by the office of Research and the University Graduate School at Indiana University in the form of a Summer Faculty Fellowship that was crucial to the early stages of this project, and by the Dean of Faculties office, which funded a yearlong Multidisciplinary Seminar on systems theory and postmodernism that I codirected.

In this connection and in others, I owe a very special thank you to my friend and collaborator Bill Rasch.

INTRODUCTION

Nothing Fails like Success

The Postmodern Moment and the
Problem of the "Outside"

THIS STUDY locates itself in the wake of what has often been characterized as the "crisis" of postmodern theory, a crisis brought about by what Jean-François Lyotard, Michel Foucault, Jacques Derrida, Richard Rorty, Gilles Deleuze, and other leading theorists of "the postmodern condition" have characterized (to use Lyotard's phrase) as an "incredulity toward metanarratives." According to Lyotard, whose work on the postmodern may be taken as exemplary, this crisis is twofold. First, "to the obsolescence of the metanarrative apparatus of legitimation corresponds, most notably, the crisis of metaphysical philosophy"—that is, of the traditional philosophical and critical paradigms of the Enlightenment and of the modern period generally (subject versus object, culture versus nature, organism versus environment, spirit versus matter), which have historically enabled philosophy and cultural critique of either the realist/materialist (Marx) or idealist (Descartes, Kant, Hegel) variety.[1] Second—and perhaps more important, depending on your view of the proper relationship of philosophy and political practice—this crisis is anything but merely theoretical, for, as Lyotard points out, the traditional philosophical paradigms and the metanarratives they make possible provided the foundation for the *political* projects of modernity, which base themselves on "the progressive emancipation of reason and freedom," whether in the form of historical materialism, parliamentary liberalism, or in other ways.[2]

In his characterization of the double crisis of postmodernism, Lyotard puts his finger on what is perhaps *the* central political challenge for contemporary intellectuals who are interested in doing socially and politically responsive work *as* intellectuals, but who have discovered over the past two decades that nothing fails, so to speak, like success. What I mean by this is that contemporary intellectuals have found themselves faced with the following conundrum: On the one hand, the critiques of the traditional philosophical paradigms of positivism, empiricism, and the like, which stress instead the contingency and social construction of knowledge (pragmatism, poststructuralism, materialist feminism), would seem politically promising because they hold out hope that a world contingently constructed might also be *differently* (i.e., more justly) constructed. But, on the other hand, that very constructivist account has left intellectuals seeking grounds for their own political practice without a foundation from which to assert the privilege of their own positions. Having undercut the philosophical footing of those in power, contemporary intellectuals find their own supposedly more progressive claims in danger as well of being "just another" contingent (and, from a cynical point of view, self-serving) interpretation.[3]

On one side, then, we find critics of diverse political stripe who lament that the breakdown of the realist philosophical world worldview means a loss of reference and meaning that undermines the ethical and political promises of Enlightenment modernity. Defenders of the realist tradition hold, to put it schematically, that interpretive validity depends on the representational adequation—the faithful mirroring, as Richard Rorty has argued[4]—of the objective meaning of the text, the event, or the social phenomenon. From this perspective, if objectivity or something very much like it is not possible, then we are automatically driven back upon relativism and even nihilism. On the other side, we find proponents of postmodernism such as Lyotard, Rorty, and Foucault who celebrate this very loss of representational authority as a liberation of the social and cultural field from what Jacques Derrida has famously called "logocentrism," a liberation that returns interpretive activity to the materiality, historicity, and social embeddedness of its processes and practices of production. To which critics of postmodernism respond in turn that these theorists cannot claim that such a breakdown of realism has taken place without engaging in a self-refuting paradox; as one recent study puts it, "How does one rule out categorical theories in principle without getting categorical? How does one universalize about theory's inability to universalize?"[5]

As I will argue in more detail later, this rather widespread charge against postmodern "relativism" has been convincingly refuted, to my mind, by all three of the theoretical approaches that converge in these pages: pragmatism, systems theory, and poststructuralism. Richard Rorty's response from within the pragmatist camp is especially lucid:

> the pragmatist is not holding a positive theory of truth which says that something is relative to something else. He is, instead, making the purely *negative* point that we should drop the traditional distinction between knowledge and opinion, construed as the distinction between truth as correspondence to reality and truth as a commendatory term for well-justified beliefs.... [T]he pragmatist does not have a theory of truth, much less a relativistic one.[6]

As Rorty suggests, the charges of "relativism," "self-refutation," and "performative contradiction" often made against postmodern constructivist theories miss the point, because they fail to recognize the inadequacy of the terms that frame the dispute itself. What they overlook is that for *both* realism (and its extreme form, positivism) *and* idealism (and its extreme form, relativism), " 'making true' and 'representing' are reciprocal relations: the nonlinguistic item which makes S true is the one represented by S," with the realist believing that "inquiry is a matter of finding out the nature of something which lies outside the web of beliefs and desires," and the idealist holding that one can "derive the object's determinacy and structure from that of the subject" or, in more contemporary versions, from "mind" or "language" or "interpretive communities" or "social practice" (*ORT* 4, 96, 5). The problem with both of these positions, and with the representationalist frame in general — and here postmodern theory would seem to coincide for once with our commonsensical intuitions — is that

> neither does thought determine reality nor, in the sense intended by the realist, does reality determine thought. More precisely, it is no truer that "atoms are what they are because we use 'atom' as we do" than that "we use 'atom' as we do because atoms are as they are." *Both* of these claims, the antirepresentationalist says, are entirely empty. (*ORT* 5)

In the meantime, most contemporary critics have settled for an uneasy compromise somewhere between these two poles, out of the line of fire, as it were, believing (if only tacitly) that there is indeed a preexistent, finite reality with its own independent nature, but which is viewed differently by different observers according to the cultural and social determinations that shape their vision of things. The problem with this commonsensical view, however, is that it is purchased at the expense of extreme incoherence, since it endorses the very representationalism that it claims to disavow; that is, if that preexisting objective reality *is* viewed differently by different observers, then how can one know that it is, indeed, objective, much less measure the differences between these views?

If Rorty is right—and I think he is—the worst thing intellectuals can do is to retreat, even under the spur of progressive political intentions, to the very foundationalist notions that they themselves have called into question. To do so would be to admit, in so many words, that theory is always already self-serving anyway, which is to say, in a sense, that theory does not exist at all, that there is nothing but practice—a putatively (but only putatively, as we shall see) "pragmatist" claim that we will find wanting in our discussion in chapter 1 of Walter Benn Michaels and the "Against Theory" polemics of the 1980s. My guiding conviction here, then, is that theoretically and politically, the only way out is through. The representationalist stance has played itself out, resulting in a series of stale and unproductive debates, and it must be abandoned, I believe, in favor of a broad and thoroughgoing theoretical pragmatism—but a pragmatism that does not engage (as the Rortyan variety finally does) in what Cornel West calls an "evasion" of the epistemological challenges raised by postmodern theory.[7] And it is doubly important that a renovated pragmatism confront these challenges because the most immediate danger for any pragmatist philosophy is the specter of philosophical idealism. As Rorty himself recognizes, if "we can only inquire after things under a description," then this immediately raises the suspicion that antirepresentationalism is simply transcendental idealism in linguistic disguise...one more version of the Kantian attempt to derive the object's determinacy and structure from that of the subject" (*ORT* 100, 4)—or, in its more contemporary variants, from the language games and discourses of specific interpretive communities.

Here again, nothing fails like success, for one of the central ironies of postmodern theoretical discourse is that materialism flip-flops over into idealism precisely because the social constructionist account of knowledge has been so thoroughly taken for granted since the epochal work of poststructuralism in the 1970s, and, beyond that, since the "linguistic turn" in twentieth-century philosophy that we find in Wittgenstein, Searle, Rorty, and others. More and more, it seems, the social constructionist critique of transcendence and the metaphysics of presence (which was supposed to return meaning and interpretation to the social, historical, and material processes of their production) has turned instead into its *own* form of idealism, one that often behaves as if what used to be called the "referent" or "object" of knowledge—the "outside" of any given discourse, to use Deleuze and Foucault's phrase — is *nothing but* what a particular discourse makes of it. What has happened, in other words, is that what started out as a liberating postmodern heterodoxy has stiffened into its own form of orthodoxy.

Michael Taussig, for example, has complained about what he calls the "epistemically correct" and "once-unsettling observation that most of what seems important in life is made up and is neither more (nor less) than, as a certain turn of phrase would have it, 'a social construction.'" As Taussig puts it, when theory came to understand that "race or gender or nation . . . were so many social constructions, inventions, and representations, a window was opened, an invitation to begin the critical project of analysis and cultural reconstruction was offered." But "what was nothing more than an invitation, a preamble to investigation has, by and large, been converted instead into a conclusion — e.g. 'sex is a social construction,' 'race is a social construction,' 'the nation is an invention,' and so forth, the tradition of invention."[8]

Taussig's invocation of "the tradition of invention" is entirely to the point, for, as Brian Massumi suggests, the hegemony of this sort of social constructionism has ironically created a situation in which "the classical definition of the human as the rational animal returns in a new permutation: the human as the chattering animal." What started out as a revisionist theoretical program devoted to breaking down logocentrism and the last vestiges of humanism has instead wound up reinstating "a rigid divide between the human and the nonhuman" that leads to a pervasive "cultural solipsism." So it is that "theoretical moves aimed at ending the Human end up making human culture the measure and meaning of all things, in a kind of unfettered anthropomorphism."[9]

If the suspicions voiced by Taussig and Massumi are justified — and I think they are — then my investigations here may be said to locate themselves not so much in the "crisis" of postmodern theory, but more specifically in *the crisis of that crisis*, one that suggests that postmodern theory needs to renew its commitment to theoretical *heterodoxy* by confronting its own *orthodoxy* with the problem that lends its name to my title: the problem of the "outside" of theory. Hence, the double imperative of this study, the first more theoretical and indeed epistemological in focus, the second more explicitly material and political: first, to explore how the varieties of postmodern theory examined here (pragmatism, poststructuralism, and systems theory) confront the specter of philosophical idealism and the "unfettered anthropomorphism" that perpetuates it in theorizing their relation to an "outside," an object, or, if you like, a "real world" not wholly constituted by discourses, language games, and interpretive communities; and second, to assess those confrontations in light of an essentially *pragmatic* view of theory, one that constantly asks what practical and material difference it makes, and to whom, how these fundamental epistemological problems are negotiated.

For my purposes here, "pragmatism" may be characterized by two main features: first, in epistemological terms, its resolute antifoundationalist and antirepresentationalist stance, which eschews philosophy as a mode of "transcendental inquiry"; and second, its relative instrumentalism and commitment to foregrounding the practical, material effects of thinking, its interest in what James called "the cash value of thought."[10] Pragmatism is also distinguished—not only in Emerson, James, and Rorty, but also in Deleuze and in Maturana and Varela—by its integrationist and contextualist rather than atomistic and analytical approach, one that holds that experience is rendered meaningful and coherent by organizing structures, patterns, gestalts, or language games that are themselves denied any foundational ontological status. Hence—and again this links both systems theory and Deleuzian poststructuralism rather directly with the philosophy of James and Peirce—pragmatism holds a particular theory of truth: an operationalism for which "Truth is 'the successful working of an idea' within a specific (and always limited) context. Truth is verification in practice."[11] In view of the pragmatist impulse that stretches from James to the poststructuralism of Deleuze, the function of philosophy and theory is thus the creation of new concepts whose value is to be judged largely by their effects in a whole range of contexts.

We can bring the specificity of pragmatism into even sharper focus by noting that its emphasis on theory's instrumentalism is often at the same time a devaluation of epistemology for its own sake—a tendency that is clear not only in James's definition of "the pragmatic method" (and in the conscientiously homely examples he uses to illustrate his point),[12] but also in Walter Benn Michaels's "Against Theory" polemic, in essays of Rorty's such as "The Priority of Democracy to Philosophy," and, in a different way, in Stanley Cavell's reading of "moral perfectionism" out of Emerson, which seeks to preserve philosophy, to be sure, but very much in the pragmatist tradition of *anti*philosophy, with the philosopher (as Cavell puts it in a winning phrase) as "the hobo of thought."[13]

These commitments are shared, of course, by other sorts of theory that do not call themselves "pragmatist." Philosophical antifoundationalism, for example, is perhaps most strongly associated with deconstruction, and an interest in the instrumentalism of theory has traditionally been philosophical terra firma for Marxism. But it is not too much to say that pragmatism is more resolutely committed to both of these priorities than either of its more illustrious rivals. On the first point, pragmatism's antifoundationalism may be seen as more thoroughgoing than deconstruction's because, as many critics have noted,[14] it sees knowledge as produced by what Hilary Putnam has characterized as a "Kantian pluralism" of language games, conventions, and discourses,[15] whereas deconstruction is fond of making general claims about the nature of thinking, language, or writing as such. Deconstruction, that is, is antifoundationalist to be sure, but, we might say, in the mode (as Pierre Bourdieu has reminded us) of foundationalism.[16] As Cavell has remarked, such claims in deconstruction often "seem as self-imposed as the grandest philosophy — or, as Heidegger might almost have put it, as unself-imposed."[17] And as for the instrumentality of philosophy, it could be argued that pragmatism is in fact more committed in this regard than Marxism, simply because it is open to a wider range of instrumentali*ties* than Marxism, which has typically maintained that commitment in the more narrow terms of the class struggle or the economic as such.

I have already touched on the difficulties experienced by the American pragmatist tradition in its attempt to make good on this double imperative, difficulties that stem in large part from pragmatism's characteristic posture (stated or unstated) that (to use Rorty's phrase) "we already have enough theory." As we shall see in chapter 1, pragmatism in Rorty, Michaels, and Cavell brings front and center the revisable, self-critical, and reflexive nature of all beliefs and descriptions, but only to recontain that commitment to contingency and the incipient pluralism it promises within an ideology of liberalism that, in its Rortyan version, declares out of the picture from the outset those social others whose very otherness or difference might lead to the critical reassessment of the beliefs of the liberal *ethnos*. Hence, these versions of mainstream American pragmatism give us no way to theorize the *productive* and *necessary* relationship between antagonistic beliefs in the social sphere. It is on the terrain of this last problem that both poststructuralism and systems theory will take a decisive step beyond mainline American pragmatism — a step predicated on the understanding that a philosophical commitment to theorizing the pragmatics of contingency needs *more* epistemology-centered philosophy, not less.

As we will see in chapter 2, the priority of systems theory resides in its pursuit, rather than "evasion," of the problem of the contingency of knowledge — a problem from which systems theory will attempt to derive a thoroughgoing theoretical pluralism. In contemporary systems theory, the problems of circularity, self-reference, and the unpredictable effects of recursivity serve as the keystone, rather than the bête noire, for a pluralist theory of interpretation and observation. Like Rorty, Niklas Luhmann stresses the contingency of interpretation (or "observation," to use his term), but then takes a crucial additional step in arguing that all observations are based on a constitutive distinction (between figure and ground, say, or legal and illegal) that is paradoxical because it posits the identity of difference (the distinction between legal and illegal is itself made within the legal, i.e., within one side of the distinction). For Luhmann as for Humberto Maturana and Francisco Varela, the only way to cut the "Gordian knot" of the realism/idealism debate is to follow through to its conclusion the problem of contingency, to assert that "everything that is said is said by someone," and to *then* remember that all such assertions are based on a "blind spot" of paradoxical distinction that not the observer in question, but only *other* observers, can disclose (one cannot acknowledge the paradoxical identity of legal and illegal, for example, and at the same time operate within the legal system; only *another* observation, made from another system, can make such a critical observation). Self-critical reflection is thus, strictly speaking, impossible, and must instead be distributed in the social field among what Luhmann calls a "plurality" of observers. Thus, Luhmann—contra Rorty—derives from the epistemologically tautological and self-referential status of any observation the *necessity* of the observations of others, thus installing the epistemological conditions of possibility for an incipient pluralism at the heart of the foreclosed Rortyan "we."

Here, systems theory's insistence on the constitutive "blind spot" and plurality of observation bears comparison with the theory, in Ernesto Laclau, Chantal Mouffe, and Slavoj Žižek, of an irreducible "antagonism" derived from the "nonsutured" character of the social. These theorists, like Luhmann, do not disavow or repress what Žižek calls the "broken and perverted" nature of communication, but instead attempt to derive from it the conditions of possibility for democratic sociality. Like the theorists of social antagonism, Luhmann insists that such "blockages," "deadlocks," or aporias do not impede but rather *make possible* a pluralist society; hence, a truly pluralist philosophy must be *post*modernist in the sense that it must avoid at all costs the quintessentially modernist and Enlightenment strategy of reducing complexity in the name of social consensus.

To provide an introduction for American readers to systems theory (the first time for most, I expect) and give it pride of place alongside the more familiar theories of pragmatism, poststructuralism, and post-Marxism is an important if secondary aim of my study—an ironic belatedness in itself, given that systems theory has its roots in American soil.[18] But an even more important reason I include it here is that the epistemological problems vigorously engaged by systems theory across disciplinary lines (in biology, sociology, information engineering, and much else) have, in the humanities, been typically posed as problems of language or textuality. It is precisely here, I think, that we should remember the sorts of admonitions about facile constructivism that we find in Taussig and Massumi, and remind ourselves how often humanist theory has simplified itself—purified itself, as it were—by positing a privileged relation of the human to either the presence or the absence of language, the signifier, the phallus, the soul, reason, toolmaking, and so on. It is here that attention to the encounters with the "outside" of theory in areas in like cognitive science (instead of literary theory) and under the paradigm of "observation" (instead of interpretation) might prove useful in confronting the *human* sciences with a *disciplinary* "outside" that might help reveal some of the humanities' underexamined assumptions and procedures.

In this light, we need to keep in mind that the "outside" of my title refers not to ecology in the usual sense nor to "the Real" of psychoanalysis, but rather to one side of the system/environment distinction, a distinction utilized not *only* by systems that are either language- or text-based—that is to say, not only by systems that are either human and/or human*ist*. This seems to me an especially distinctive and promising feature of systems theory, one that might more readily engage the "hybrid" or "cyborg" networks of postmodernity (compellingly theorized by Bruno Latour, Donna Haraway, and others), which include all sorts of nonhuman agents and actors—a challenge to which the old ontological dualisms of subject/object, organism/machine, and so on would seem to be woefully inadequate.[19] This crucial *posthumanist* dimension suggests the priority of systems theory not only over deconstruction for "new social movements" such as ecology and animal rights, but also over the theory of social antagonism as we find it in Žižek, which remains ineluctably tied to the figure of the Human and the Oedipal (even if it reverses humanism's ethical valences).

As we will see, a similar reluctance to base theory on the textual, linguistic, or semiotic model distinguishes the work of Deleuze and Foucault from other versions of poststructuralism. At the same time, however, a signal difference between systems theory and poststructuralism is that the Achilles' heel of the former has so far been its lack of a coherent account of its own ethical and political implications, about which even its main practitioners (Maturana and Varela on the one hand, and Luhmann on the other) would seem to be in utter disagreement, with Luhmann often endorsing what amounts to a liberal technocratic functionalism not very different from Rorty's own, and Maturana and Varela espousing a suspiciously humanist ethics that seems completely at odds with their posthumanist epistemological innovations. And even if we do not (and I think we should not) agree with the garden variety ideological critiques of systems theory—that it is, as Peter Galison puts it, "the apotheosis of behaviorism," which makes "an angel of control and a devil of disorder"[20]—we are nevertheless forced to conclude that a serious short-coming for systems theory has been its inability or unwillingness to confront the problems of *power* and social inequality that belie its theory of the formal equivalence and contingency of all observation, and often render such equivalence beside the point; for, as Donna Haraway rightly reminds us, observation "is always a question of the power to see."[21]

A commitment to confronting the dynamics of power and its relation to multiplicity and difference is everywhere present, of course, in the work of Foucault and Deleuze, and it is that unstinting interest that leads me to read them as exemplary poststructuralists for the pragmatist orientation of this study. In my view, Deleuze and Foucault not only may but *must* be read as distinctly postmodern pragmatists who seek to theorize the relation between contingency, the "aleatory Outside," and what Deleuze finds in Foucault: the possibility of "new coordinates for praxis." Once we have dispensed with Rorty's (mis)reading of Foucault—which hinges in no small part on its failure to understand the importance of what Foucault characterizes as the "productivity" of power and the materiality of practice—we can better see what joins Foucault with Deleuze: a commitment to an "ethics of thought" that places a premium on the production of new concepts by means of the continual confrontation of thought with its own outside.

And here, precisely, is where the prying open of pragmatism by systems theory via a renewed interest in epistemological problems like "the observation of observation" is joined not only by the theory of social antagonism, but also by the work of Deleuze, which provides what is finally an ontology rather than an epistemology of the conditions of possibility for democratic pluralism. As in systems theory's vision of the distribution of observation in a horizontal, functionally differentiated social space, Deleuze's work, as Michael Hardt suggests, helps us "develop a dynamic conception of democratic society as open, horizontal, and collective," as "a continual process of composition and decomposition through social encounters on an immanent field of forces."[22] As we shall see, the aim of Deleuze's metaphysics is not to discover a resting place for thought or existence, but rather to open up this field of forces to analysis toward thoroughly pragmatic ends. The political dimension or "relevance" of Deleuze's thought, which often seems oblique, resides in no small part in its refusal to see its vocation as providing "grounds" or "frames" or "foundations" for a particular practice. Deleuze's thinking is concerned instead with the conditions and dynamics under which specific forms of resistance are possible in the ongoing struggle between hegemonic social cartographies or "diagrams" and their own outsides.

Such analysis is of immense pragmatic importance in addressing the "new social movements" (environmentalism, sexual minorities, and so on) that traditional Marxism has often discounted as "epiphenomenal" or "diversionary." For Deleuze, what the events of May 1968 in France demonstrated was the inability of traditional frames of the theory/praxis relation to understand that the truly revolutionary political potential of the moment lay beyond the strict domain of the class contradictions of capitalism. What is invaluable for pragmatist theory about Deleuze's work, in other words, is its recognition of the crucial *micropolitical dimension* of capitalist culture—a recognition shared even more pronouncedly in Foucault's articulation of the relation between power and knowledge through his analysis of the disciplines. Foucault's anatomy of this "counter-law" at work in social formations emphasizes the *materiality* of practice, what we might think of as the materialist "unconscious" of social and (though Foucault would not use the term) ideological reproduction.

Although Foucault's work on the disciplines and on "panoptical" society more generally is of immense importance, what is less well known is a dimension of his thought that links him to the Deleuze of "forces" and "lines of flight" and, beyond that, to the "anarchistic" side of William James and the "whimsical" Emerson of essays such as "Self-Reliance." This is the sense of pragmatism, as we shall see, foregrounded by Stanley Cavell's vision of philosophy (and of Emerson) as a task of "transience" and "onwardness," a process that is crucial to the project of "moral perfectionism" and, in his view, to democracy. Similarly, what Foucault characterizes as the "ethics of thought" is a "constant 'civil disobedience' within our constituted experience," as John Rajchman characterizes it, one that "directs our attention to the very concrete freedom of writing, thinking and living in a permanent questioning of those systems of thought and problematic forms of experience in which we find ourselves."[23]

These two strands of Foucault's pragmatist thinking are conjugated with remarkable insight and originality in Deleuze's book on Foucault, not least in what Deleuze theorizes as "the fold"—a difficult and ambitious figure that attempts, through a topographical treatment, to make good on the impulse at work in systems theory: to see the outside not as a naturally given ground or totality, but as the outside *of* the inside. Unlike systems theory's handling of the problem, however, Deleuze's fold crucially reverses this orientation and pursues the inside as "the inside *of* the outside," a reorientation that is symptomatic of Deleuze's final commitment to ontology and the univocity of being, rather than (as in systems theory) to epistemology and difference. The Deleuzian fold would suture closed with ontological substance, as it were, the open space or vacuum between points, observations, and, finally, between the inside and outside that systems theory attempts to leave open.

I conclude by framing my own view of the relationship between politics and theory in light of how (post-)Marxist theory has negotiated postmodernism's "constructivist revolution." Here, I pay particular attention to the work of one of our most exemplary and politically engaged theorists, Fredric Jameson. As we will see, I share Jameson's commitment to the necessity of a broadly operative anticapitalist politics, but I resist his grounding of that political practice in the imperative of totalization and the dialectic—an imperative that, even as it insists on the unmappability of the outside under postmodernism, seeks to occupy an authoritative space *outside* that outside, a theoretical purchase from which that first outside can and must be seen as expressive of a unitary cause: global capitalism. In my view, no such space is available, and so the political commitments and claims that I share with Jameson can only be made *pragmatically*. To use Luhmann's systems theory vocabulary, there *is* no distinction—including the Marxist distinction of the socially determinative priority of the economic mode of production—that is a final, noncontingent distinction. What this means is that theory cannot provide a *grounding* for politics and praxis in the way that Jameson imagines.

We must opt instead, I argue, for what Kenneth Burke calls a "comic perspective" on the relationship between a theoretical commitment to contingency, difference, and "permanent critique" on the one hand, and a political commitment to material and social praxis on the other, with each serving as the other's "bad conscience" in a ceaseless, democratically productive antagonism. The comic frame, according to Burke, "considers human life as a project in 'composition,'" one that offers "maximum opportunity for the resources of *criticism*"; it should "enable people *to be observers of themselves, while acting*," and push the thinking subject to "'transcend' himself by noting his own foibles."[24] The comic frame does not provide a "ground" or "foundation" for praxis but only "damage control" for praxis, which is *always* reductive of difference (or, in systems theory language, of an outside environment that is always already more complex than the system itself). But the Burkean "comic" attitude in and of itself, of course, is not enough, because expressing the desirability of open-mindedness or self-criticalness is not, by a long shot, the same as *having a rigorous and coherent theoretical account of that desirability's necessity*. Whether or not the "comic attitude" constitutes a distinctly "postmodern" solution to the relationship between theory and politics—and how that solution relates to the problem of increasingly globalized capitalism—is an issue on which major theorists such as Jameson and Luhmann disagree. But we need both, I think—and their disagreement—to provide a theoretically compelling and politically responsive account of our contemporary situation.

The aim of this project, then, is twofold: first and most important, to explore the theoretical, political, and ethical dimensions of how some of the major theorists within "postmodernism" have confronted the problem of thinking the "outside" of theory; and second and subsidiary, to place alongside those investigations lesser-known but immensely promising developments in the sciences and social sciences in systems theory that provide perhaps the most compelling models yet constructed for dealing with these complexities. My unabashedly abstract and theoretical approach to these issues raises an obvious question: Does my interest in what Rodolphe Gasché characterizes as the "infrastructural" theoretical difficulties common to all these efforts imply that all such problems of the outside—of sex, of production, of ecology, of animal rights—finally reduce to the *same* problem? The answer, I believe, is "yes and no." "Yes" in the sense that any attempt to elucidate the problematics of any of these issues—much less ground their privileged status for cultural studies and social theory—will unavoidably have to grapple with the general epistemological and theoretical challenges that form the focus of my investigations. Nothing less than the credibility of political-intellectual work, *as* intellectual work, is at stake when we level against the forces of patriarchy, whiteness, and economic privilege the deconstructive critique of logocentrism and the postmodern demolition of foundationalism, and then epistemologically look the other way when attempting to ground a politics and praxis we support. As Barbara Herrnstein Smith has lucidly observed, it is not as if "objectivism is wrong when practiced by the wrong people for the wrong reasons, but right when practiced by the right people for the right reasons."[25] In that sense, all these problems *are* the same problem, insofar as they all *have* the same problems; that is why they have much to teach each other.

But in a different way—in a way underscored by my emphasis on the contingency and *pragmatics* of theory—all of these problems *do not* boil down to the same problem, because how one confronts the theoretical challenges I have just sketched will be different at different times and in different contexts for the feminist, or the labor organizer, or the environmentalist, or the gay rights activist. Pragmatically speaking, there are times—as with feminism's appeal to the solidarity of global women's experience, or with animal liberation's reliance on a liberal humanist rhetoric of "rights"—when, for strategic reasons, one will want to mobilize very problem-specific and audience-oriented rhetorical strategies that avoid the embarrassing theoretical problem of their own contingency. But that is always a dangerous business, I think, simply because history provides too many examples of such sleight of hand in the name of political expediency getting *out* of hand. There may indeed be times when the rigors and self-scrutiny of theory can be shelved for the sake of politics and practice. But this — here, in this book — is not one of those times.

O N E

Pragmatism

Rorty, Cavell, and Others

IN THIS chapter, I examine a range of theorists who have been associated with "pragmatism" or "neopragmatism," a critical genealogy that stretches back to the Ralph
Waldo Emerson of the 1830s, then to William James—who may be said to have
given the theory (or, more properly, the antitheory) its name (which he adapted
from the "pragmaticism" of Harvard colleague Charles Sanders Peirce)—then to
John Dewey, and reaching forward finally to contemporary critics and philosophers
such as Frank Lentricchia, Cornel West, and the figures I will examine in most detail: Walter Benn Michaels, Richard Rorty, and Stanley Cavell. These three represent the range of contemporary pragmatism while still pursuing an identifiably
common set of concerns under the same broad set of theoretical assumptions. As
we shall see, my main contention about the politics of pragmatism as practiced by
these three is that, in all of them (and this despite their differences), an identifiably
liberal problematic that cannot take account of asymmetries of power in the social
field operates essentially, if sometimes obliquely, as what Fredric Jameson has called
an ideological "strategy of containment" that undermines or short-circuits what could
otherwise be viewed as a Nietzschean (and, via the lines of influence, Emersonian)
pragmatist commitment to radical plurality, contingency, historicity, and difference.[1]

The central pragmatist contention that joins Walter Benn Michaels
and Richard Rorty reaches back to the founding contention of James himself: that,

for the pragmatist, "truth" means "what it is better for *us* to believe," it is what is "good in the way of belief" (Rorty), and so "Meaning is not filtered through what we believe, it is constituted by what we believe" (Michaels). As we shall see, however, it is important to distinguish Rorty's account of belief and what falls outside it from that of Michaels (and his former teacher, Stanley Fish) and the "Against Theory" position—and this despite Rorty's professed endorsement of the Michaels/Fish line in the essay "Texts and Lumps." Rorty's conjugation of the relationship between belief and theory (reflection on belief) must be distinguished from the more idealist treatment we find in the Michaels/Fish line. In this light, Stanley Cavell, whom I will take up last, looks like odd man out (a happenstance with which he would, I think, be pleased), for what separates Cavell from both Rorty and Michaels is his ongoing engagement with the problem of philosophical skepticism—a problem that is simply a nonissue, it is safe to say, for Michaels and Rorty. Indeed, if what distinguishes Rorty's pragmatism is its reactivation of an entire philosophical style and tradition stretching back to Emerson and forward through Wittgenstein, and what marks Michaels's is its polemical "antitheory" stance, then what makes Cavell's version of pragmatism unique is its attempt to combine the *desire* for the "outside" of theory and philosophy (which skepticism keeps alive as it "mourns the passing of the world") with a commitment to antifoundationalism and contingency, to philosophy, in Cavell's words, as a task of "onwardness," "transience," and "homelessness," to thinking as "finding" rather than the "founding" of foundational philosophy.

The Island of Belief: Walter Benn Michaels and the Uses of William James

The problem of "belief," Walter Benn Michaels noted early in his career, is "one of the few problems in literary theory which Anglo-American critics can with some justice claim as our own."[2] In "Saving the Text" (1978), Michaels offers a concise definition of belief that will carry through to his later essays "against theory" and his important New Historicist study *The Gold Standard and the Logic of Naturalism:*

> These solutions [to the problems posed by our inability to achieve "disinterestedness" in our criticism] are in many ways very different but the view that they have in common is that our beliefs are like filters through which we more or less accurately discern texts—the optimists imagine these filters growing ever more transparent, the pessimists ever more opaque. What I should like to suggest here is that both these view are mistaken because the model they hold in common is mistaken. Our beliefs are not obstacles

between us and meaning, they are what make meaning possible in the first place. Meaning is not filtered through what we believe, it is constituted by what we believe. (780)

Michaels's critique of "distinterestedness" is certainly salutary when taken on its own. The problem, of course, is that we cannot take it that way, because Michaels extends (and overextends) it with Steven Knapp into a full-blown critique of "theory" in his later work. Let us leave aside for the moment Knapp and Michaels's rather peevish (and, in Fredric Jameson's words, "reassuringly restricted") characterization of "theory" as "a special project in literary criticism."[3] Instead, I want to notice what the polemical brouhaha over the "Against Theory" project obscures: that Michaels's critique of "distinterestedness" on behalf of "belief" would seem to promise a pragmatist micropolitical analysis of the institutional production of belief on the model of Gerald Graff's *Professing Literature*, Richard Ohmann's *English in America*, or Barbara Herrnstein Smith's *Contingencies of Value*. And this promise seems only extended in Michaels's contention later in "Saving the Text" that his position provides a way into "an objectivity that is limited but real," one "based not on the attempt to match interpretations up to a text that exists independently of them, but based instead on what readers believe" (787–88).

That promise, however, will remain unfulfilled as Michaels fleshes out the general epistemological structure of "belief" and extends the claims for its seamlessness and all-constituting power in ways that will undermine what initially seems most compelling about his account: that all interpretive choices, even when they seem "free," are reproductive of previously held beliefs that the subject cannot, through critical reflection, fully master, even though she may now want to abandon them. In "Is There a Politics of Interpretation?" for example, Michaels's aim is to demonstrate the quintessentially theoretical claim (and this hard on the heels of "Against Theory") that "it does not make sense to say that you choose to believe anything at all."[4] This is so, he argues, because the epistemological freedom required by the category of "choice" is fundamentally at odds with the epistemological compulsion named by the category of "belief." Michaels's version of the paradox goes like this: if you are free enough from assumptions and beliefs to make a choice that is truly a *free* choice, then you are by that same logic unable to make any choice at all because you will have no criteria on which to base that choice. Conversely, if you *do* possess the necessary criteria to make such a choice, then it will no longer be a free choice at all, but rather an action compelled and produced by those beliefs and assumptions that provided the criteria for choosing in the first place (341, 343).[5]

This apparent frontal assault on ethical criticism (and therefore, presumably, on the liberal humanist subject and his meliorative judgment) is not only sustained but in fact intensified in *The Gold Standard*'s contention that the identity of the subject of naturalism "consists *only* in the beliefs and desires made available by the naturalist logic — which is not produced by the naturalist subject but rather is the condition of his existence."[6] And "the naturalist logic," it turns out, is the same logic that constitutes the economic totality called "the market," which, in its all-constituting power, resists all attempts to ameliorate or temper it. In fact (as practiced readers of New Historicism will have already guessed), Michaels suggests that such attempts only serve to siphon off or neutralize potentially explosive (perhaps even revolutionary) desire and discontent, thereby further reinforcing the dominance of the market and extending its logic even more insidiously into incompletely colonized enclaves of social life. In Michaels's reading, the subject is not what makes the market and its fundamental structures possible (the exchange principle, for instance), but is rather an *effect* and expression of the market. And this leads Michaels, in turn, to suggest that we abandon the concept of ideology as a critical tool and replace it with the concept of "belief" (a suggestion we will take up in more detail shortly).

At first glance, Michaels would seem to temper this sweeping claim for the all-constituting power of "belief" (and, behind it, of this thing called "the naturalist logic") by introducing what looks like an important distinction between "beliefs" and "desires." The latter — a term of considerable micropolitical resonance in the context of poststructuralism[7] — might seem to hold out some promise to trouble and destabilize the seamless social totality, but in fact that possibility is immediately foreclosed in *The Gold Standard*. In Michaels's reading of "desire," the self is constituted as a fundamental instability, a "double identity" or "internal difference" (22) generated by the market and its fundamental logics of property and exchange. The subject of naturalism, in other words, can know no completion or self-identity because it is constituted by difference.

It is this internal difference that sets going the "logic of naturalism" by which the self seeks to escape the market and the ceaselessly self-reproducing play of exchange by clinging "to definitions of texts, selves, or money," in Evan Carton's words, "as stable and essential quantities."[8] The fundamental instability of the market creates a self who therefore has, as Michaels puts it, "an insatiable appetite for representation" (*GS* 19), which manifests itself in the belief that gold is the site of natural economic value, the text is the site of stable inherent meaning apprehended by the critic's adequated critique, and the subject is the site of inalienable self-possession and free self-proprietorship.

But, in Michaels's view, this unstable desire, far from destabilizing the system, only serves further to perpetuate it, because desire is "not subversive of the capitalist economy but constitutive of its power" (48). As Fredric Jameson points out, desire for Michaels is trapped in a logic of "infinite 'supplementarity'" (*Postmodernism* 202–3); it is part of that ruse of the commodity which, in Frank Lentricchia's words, turns "the potentially revolutionary force of desire produced on capitalist terrain toward the work of conserving and perpetuating consumer capitalism."[9] To put it another way, "desire," like "belief," offers no means in Michaels's critique by which the self might be anything *other than* a purely reproductive agent of the market and its logic. It is clear from this vantage why the promise of a pragmatist micropolitics, more than hinted at in "Saving the Text," will remain unfulfilled in Michaels's later work: there is simply nothing for it to do.

But if the lack that is desire is disarmed and recontained by Michaels's critique, the political efficacy of that plenitude known as "critical distance" and "reflection" is rejected as well by definition in Michaels's concept of "belief." Indeed, what makes Michaels's "belief" what it is — in contrast to the "interests" lucidly promoted and aligned by "ideological" critics or expunged, conversely, by "disinterested" formalist ones — is that you cannot have that sort of critical distance in relation to it at all. (If you could, you would be guilty of the for-theory position opposed by Knapp and Michaels.) We can triangulate the relationship between "desire," "belief," and "ideology" in Michaels in this way: "belief" and "ideology" may exist, in Marx's famous phrase, "to reproduce the conditions of production," but because there is in Michaels's reading no Other to the market — no disruptive agent of negation (like "desire") to pose any real threat of fundamental change to the system — Michaels's concept of "belief" cannot tell us why we should *need* to secure the reproduction of a social and economic system that cannot be threatened in the first place.

Nevertheless, in *The Gold Standard*, Michaels argues that the concept of ideology must be abandoned on behalf of this notion of "belief." The concept of ideology, Michaels tells us, assumes

> the existence of subjects complete with interests and then imagines those subjects in more or less complicated (and more or less conscious) ways selecting their beliefs about the world in order to legitimate their interests. The subject of naturalism, however — at least as I have depicted him here

—is typically unable to keep his beliefs lined up with his interests for more than two or three pages at a time, a failure that stems not from inadequate powers of concentration but from the fact that his identity as a subject consists only in the beliefs and desires made available by the naturalist logic— which is not produced by the naturalist subject but rather is the condition of his existence. (177)

For Michaels, in other words, the concept of ideology must be abandoned because it "refers," as Foucault puts it in a passage Michaels quotes approvingly, "necessarily to something of the order of a subject" (177)—the subject who constitutes, as a metaphysical point of origin, the market and the logic of naturalism, and not (what Michaels wants) the reverse.[10]

This is not the place to mount an extensive review of the many debates over what ideology is and how it operates[11] (a topic to which we will return in later chapters), but two points should be raised here in response to Michaels's critique. First, this Foucauldian critique of ideology and its "constituent subject" is perfectly correct as far as it goes, but it does not go far enough—which is to say that it goes as far as the vision of "expressive totality" of Georg Lukács's *History and Class Consciousness* and not much farther. Foucault and Michaels after him are perfectly right, it seems to me, to reject the idealism of that position, which holds, as Martin Jay has characterized it, "that a totalizer, a genetic subject, creates the totality through self-objectification."[12] But that is far from saying (as Michaels's critique implies) that "ideology" is no longer a useful or powerful concept. In fact, the concept of ideology debunked by Michaels is considerably complicated by the later Marx himself, who, it may be argued, conceives the subject not as unified and constituent of the social totality, but rather as a differential, conflicted, and above all relational sort of creature whose identity is a product of what it is *not*.[13] And after Marx, the sort of ideology and its attendant subject characterized by Michaels has been rejected by many of the most influential later Marxist theorists themselves. To take only the most famous example, Louis Althusser's enormously influential "Ideology and Ideological State Apparatuses"—which holds that "ideology represents the imaginary relationship of individuals to their real conditions of existence"— conceives the relationship (as the Lacanian terminology implies) between ideology and the subject as anything *but* a facile alignment of interests by a lucid, rationalist subject.[14] Indeed, if ideology is, in Althusser's words, "a 'lived' relation to the real," an indirect and oblique "relation of a relation"[15] (of subject to an imaginary representation, and of the imaginary representation to its real conditions of existence), then how could it be otherwise? The problem, it would appear, is not so much with the concept of ideology, but rather that Michaels uses the categories of a more or less humanist subject to argue for the abandonment of the ideological subject *tout court*. And this move, I want to argue now, is itself quintessentially ideological.

Let me draw out in more detail what I have already hinted at: that Michaels's concept of "belief" remains trapped within the conceptual limits of the liberal problematic. For Michaels's revisionary claims about the power of "belief" and the foreclosure of "choice" to have any force, his argument must continue to rely on the values and the conceptual coordinates of the liberal subject. Time and again in Michaels—as in many another version of liberalism—we find ourselves in the position of choosing either *that* sort of centered, self-reliant subject or no subject at all; either perfectly unfettered freedom of choice on the one hand, or totalizing and overstructuring "belief" on the other.

And the terrain of belief is not the only place in Michaels where the liberal politics of pragmatism plays itself out in the name of William James. A second critical topos, in *The Principles of Psychology*, is central to the "Introduction" to *The Gold Standard*, where Michaels draws our attention to the fact that James's model of selfhood is essentially a model of ownership. *The Principles of Psychology* rivets our attention on what may be viewed as the primal scene of liberalism: the conception of autonomous selfhood in accordance with the language and structure of private property. To discuss the genealogy of pragmatism, we must deal with James. And to deal with James, we must, it seems, confront his Lockean liberalism.

In *The Principles of Psychology*, James asks, what does consciousness "mean when it calls the present self the *same* with one of the past selves which it has in mind?" (quoted in *GS* 7). Our selves, James decides, must be joined by what Michaels calls "common ownership," but what or who, then, does the owning? Michaels sketches the problem for us:

> How, then, can we account for the way in which the present thought establishes ownership over past thoughts? Our mistake, James thinks, has been to imagine the thought as *establishing* ownership over past thoughts; instead, we should think of it as *already* owning them. The owner has "inherited his 'title.'" His own "birth" is always coincident with "the death of another owner"; indeed, the very existence of an owner must coincide with the coming into existence of the owned. "Each Thought is thus born an owner, and dies owned, transmitting whatever it realized as its Self to its own later proprietor." (9)

For Michaels, the Jamesian model of selfhood is a quintessential example of the totalizing power of the logic of the market; James's account of the self is, in Michaels's words, a story of "the continual transformation of owner into owned" (22). In James, we learn a lesson that *The Gold Standard* aims to teach us again and again: that the logic of the market is not the effect of selfhood but its very condition. And consequently, the Jamesian self who might critically reflect on his beliefs can do so only within the purview of the market logic that constitutes the self who might engage in such reflection.

From this vantage point, James's attempt to save the self from the brutalities of the early modern American variety of capitalism—with its Taylorization, its imperialist incursions in both Atlantic and Pacific, and its "abstraction" and "rationalization" (as James liked to call it) of cultural and intellectual life—turns out to be an unwitting repetition and indeed an insidious internalization of that very social and economic totality. Far from providing a stay against alienation and ruthless competitive individualism, James's Lockean property model of selfhood guarantees it, because freedom on this model is the right to dispose of and enjoy the property of the self, its capacities and potentialities (including, of course, its labor power), as one wishes—a freedom that cannot help but be limited and threatened, however, by other self-proprietors who are trying to achieve the same sort of freedom. All of which is simply to grant the classic Marxist critique: that insofar as the self is conceived as a kind of private property, I will alienate and threaten your freedom insofar as I realize my own, and vice versa.[16]

To more fully understand the attractions and limitations of Michaels's reading of this problem in James,[17] it might be useful to compare it with Frank Lentricchia's powerful rereading of Jamesian "belief" in general and of this moment in *The Principles of Psychology* in particular. Lentricchia focuses on a different passage in James, but it is one that, if anything, makes the point even more emphatically: "It seems," James writes, "as if the elementary psychic fact were not *thought* or *this thought* or *that thought* but *my thought*, every thought being *owned*." "*In the widest possible sense*," James continues, "*a man's self is the sum total of all that he can call his.*"[18] Like Michaels, Lentricchia recognizes that James's "overt commitment to the inalienable private property of selfhood . . . is an inscription of a contradiction at the very heart of capitalism"—namely, that property "can *be* property only if it *is* alienable," and that a self so conceived, therefore, is perforce an alienated self (816).

Unlike Michaels, however, Lentricchia's reading (in a moment that is perhaps overly generous to James) argues that "James employs the language of private property in order to describe the spiritual nature of persons and in an effort to turn the discourse of private property against itself by making that discourse literal in just this one instance: so as to preserve a human space of freedom, however interiorized, from the vicissitudes and coercions of the marketplace" (816). But the critical pressure of Lentricchia's reading of James must at the same time force us to ask: Can you do that? Can James or anyone so turn the historically and politically freighted rhetoric of capitalism to advantage, even in an act of engagement as diligent as James's? Lentricchia's critique allows us, I think, to answer both "yes" and "no": "no" in the sense that James's property model of selfhood reproduces the logic of alienation that it would subvert; but "yes" because, in Lentricchia's reading, that is not the end of the story. For Lentricchia is at pains to situate this moment of discursive complicity with the system in James in the context of a larger, historically specific project of anti-imperialism that the later James undertook in earnest on many different sites (writing letters to the newspaper, giving talks to grade-school teachers)—sometimes by seizing upon what Kenneth Burke would call the "ruling symbols" (like private property) of his day, and sometimes by doing precisely the reverse (as in his guerilla warfare, within the institutions of academic philosophy, against what he called "rationalism"). For Lentricchia, James's discursive complicity with the system is only one component of a larger project for social change, an undertaking that, being quintessentially pragmatist, is willing to use all that there is to use—including (especially) the politically powerful means of rhetorical identification.

Lentricchia's discussion in *Criticism and Social Change* of Kenneth Burke—another fellow traveler of pragmatism—will help to further underscore his differences with Michaels. Burke's concept of "dialectical rhetoric" helps us to see that James's appropriation of the dominant discourse (always risky and, for Burke, often necessary) is driven by a concept of political discourse as not "a simple negating language of rupture"—or, what amounts to the same thing, the pure reproduction of the market in Michaels's reading of James—"but a shrewd, self-conscious rhetoric that conserves as it negates" (*CSC* 33).[19] Lentricchia's critique of theoretical idealism serves to underscore the paralyzing effects of Michaels's homogenization of the social space in his reading of James's discursive complicity.[20] "To attempt to proceed in purity," Lentricchia writes, "—to reject the rhetorical strategies of capitalism and Christianity, *as if such strategies were in themselves responsible for human oppression*—to proceed with the illusion of purity is to situate oneself on the margin of history. . . . It is to exclude oneself from having any chance of making a difference for better or for worse" (*CSC* 36; emphasis in the original).

Lentricchia insists on an unruly materiality of the social that is never wholly identical with the languages that seek to produce and master it. On that view, a pragmatist critique must attend not only to the abstract, symbolic, or textual form of a concept but also to what Burke calls its "bureaucratization" in material, social form: "Pragmatism," Burke writes—in a wonderful meditation on what he calls the "unintended by-products" of abstract concepts—"would note how the particular choice of materials and methods in which to embody the ideal gives rise to conditions somewhat at variance with the spirit of the ideal,"[21] a state of affairs, it probably goes without saying, that cannot be foreseen in advance. Consequently, Lentricchia's version of Jamesian pragmatism "has no way of settling, once and for all, the question it constantly asks: Does the world rise or fall in value when any particular belief is let loose in the world?" ("Philosophers of Modernism" 805). For Lentricchia, this is James's "most unsettling insight: that a rigorous philosophy of practice and consequences cannot in advance secure consequences without establishing precisely the sort of imperial authority...which that philosophy is dedicated to undermining."[22] In fact, the stronger point is that it could never do so anyway, even if it wanted to.

  This open-endedness—this *unforeseeability*—of Jamesian pragmatism has been emphasized in other rereadings of James as well, most notably in Cornel West's *The American Evasion of Philosophy*.[23] West's critique stresses above all the fact that for James we cannot know the status of a concept or figure (the Lockean property model of selfhood, for example) until it has been let loose in the world and *only later* returned to us. West would foreground for us James's first principle of pragmatism: "There can *be* no difference anywhere," as James phrased it in a late essay, "that doesn't *make* a difference elsewhere—no difference in abstract truth that doesn't express itself in a difference in concrete fact and in conduct consequent upon that fact, imposed on somebody, somehow, somewhere, and somewhen."[24] For West, the essence of Jamesian pragmatism is its *revisability*; its first principle, in West's words, is that "the universe is incomplete, the world is still 'in the making' owing to the impact of human powers on the universe and the world" (65). For West as for Lentricchia, Jamesian pragmatism insists on the gap between concept construction on specific discursive sites and concept circulation in a broader set of contexts, and it is in that gap that the possibility of the social and the historical resides.

  All of which seems to raise for a pragmatist critique two opposite problems, which are nevertheless inextricably linked. On the one hand, this position seems to tell us that knowledge in the pragmatist sense is so deferred, dispersed, and contingent that it is hard to see how we can we reflect in any meaningful way not so much on what we think, but rather (as Michaels might phrase it) on what what we think does. On the other hand, the primacy of pragmatist agency would seem to imply that we *can* know, in a fairly direct and precise way, exactly what we are doing (hence we can, in Burkean good faith, rhetorically enlist support for it).

In fact, as is well known, James is famous for holding (though this is not the whole story on his conception of truth)[25] that truth is *"the name of whatever proves itself to be good in the way of belief, and good, too, for definite, assignable reasons"* (*Essays in Pragmatism* 156, 155; emphasis in the original).

West's critique nimbly approaches this problem by insisting on a dimension of James's thought that tends to get lost in contemporary discussions, with their emphasis on action and power. For West, as for Hilary Putnam—but not, significantly, for Richard Rorty—James's pragmatic theory of truth "preserves a realist ontology" even as it "rejects all forms of foundationalism" (67).[26] In James's words, "with some such reality any statement, in order to be counted true, must agree" (quoted in West, *The American Evasion of Philosophy* 64). At the same time, however, James stresses the constitutive and revisionist role of human action in the construction of that reality. As West puts it, "James retains a correspondence theory of truth, yet it is rather innocuous in that rational acceptability is the test for truth claims we accept" (67).

This does not exactly put to rest, however, the first problem we raised earlier: namely, how the pragmatist—who in the strong instrumentalist reading believes that critical ideas are really different only if their consequences in the world are different—can ever really ensure, through critical reflection, that her ideas will turn out to be in practice what she thinks they are in theory. Lentricchia's response to this problem is relatively straightforward, and relatively disarming: you cannot assure that, but at the same time you cannot do anything other than try to assure it. Just as Lentricchia agrees with Michaels on the *fact* (but not the interpretation) of the Jamesian property model of the self, so he reads Jamesian belief as "a 'set of rules' for the action of changing-by-interpreting the world's various texts," as "instruments of desire" of wholly "temporal character," which are "born locally in crisis and have local consequences only" (*Ariel* 106–7). But James's message for Lentricchia—and here is where he parts company with Michaels—is also that "theory" (critical reflection on "belief") is not something that one can be "for" or "against" because it is inescapable, not a matter of volition or intention but rather what James called "an appetite of the mind" and what Lentricchia calls "the need to generalize" and "to obliterate differences" (124–25). It is, to put it bluntly, a kind of conceptual imperialist within. In Lentricchia's reading of James, it is as if theory and practice are engaged in a never-ending battle on the terrain of belief.

But it seems fairly obvious that to insist, as James does, that theoretical reflection is, mutatis mutandis, an "appetite of the mind" is to reinstate once again an essentialist concept of the reflective and rational subject that is twin to the ethical, liberal self and its privileged qualities. And here, it seems to me, is where Lentricchia's reading of James is both ingenious and forthright. Rather than attempting to solve or explain away the essential liberalism of Jamesian pragmatism, Lentricchia takes it for granted. "The new pragmatists," Lentricchia writes in a passage worth quoting at length,

> flounder on one of James's strongest insights—that theory cannot be identified with agency and the self-conscious individual [and so cannot be rejected in the sense of "Against Theory"], that theory is the sort of force that tends to control individuals by speaking through them. And so does James. . . . The epistemological move to generalization may well be an "appetite of the mind" . . . [b]ut the economic and political move to generalize—the global generalization of labor known as capitalism—is not an unhistorical appetite; it is a locatable, historical phenomenon whose role tends to be blurred and repressed by James's liberal ideology of the autonomous self. (127)

What James *does* do, however, is all that a liberal discourse of individual agency and practice can do: he pushes liberalism to its absolute limits by "focusing the obsessive liberal vision of American literature at its extreme antinomian edge" where the subject dwells not as "variable expressive function of structure" but rather as "the antithesis of structure" ("Ideologies" 249). James's gamble is that he is too much within the dominant discourse of the liberal subject, but his payoff is to unleash that discourse's radically democratic tendencies against the private property side of liberalism that threatens always to recontain them.

Making Contingency Safe for Liberalism: Richard Rorty's Evasion of Philosophy

As even pragmatism's closest allies have pointed out, these two sides have been locked in increasingly pitched struggle in the work of Richard Rorty.[27] A useful staging ground for this ongoing dilemma is Rorty's vexed relationship with Nietzsche, a certain reading of whom links Rorty (as he himself has noted)[28] with the early Foucault of "Nietzsche, Genealogy, History." This is the Nietzsche who holds, as Rorty puts it, "that the philosophical tradition which stems from Plato is an attempt to avoid facing up to contingency, to escape from time and chance," the Nietzsche whose account of truth (in "Truth and Lie in an Extra-Moral Sense") Rorty quotes approvingly: "a mobile army of metaphors, metonyms, and anthropomorphisms—in short a sum of human relations, which have been enhanced, transposed, and embellished poetically and rhetorically and which after long use seem firm, canonical, and obligatory to a people" (quoted in Rorty, *ORT* 32).

My pairing of Rorty with Foucault should come as no surprise, because no two recent intellectuals have done more to call into question the philosophical tropes of vision, reflection, and their foundationalist associations. With Foucault, Rorty would seem to bring to its postmodern terminus the critical genealogy of vision and the Look that runs, in its modernist incarnation, from Freud's discourse on vision in *Civilization and Its Discontents* through Sartre's *Being and Nothingness* to Lacan's seminars and finally to influential work in the 1970s and 1980s in psychoanalysis and feminist film theory.[29] In Rorty's seminal *Philosophy and the Mirror of Nature* we find, as Cornel West puts it, "a wholesale rejection of ocular metaphors in epistemology" (202); "The picture which holds traditional philosophy captive," Rorty writes in that text,

> is that of the mind as a great mirror, containing various representations— some accurate, some not—and capable of being studied by pure, nonempirical methods. Without the notion of mind as mirror, the notion of knowledge as accuracy of representation would not have suggested itself. Without this latter notion, the strategy common to Descartes and Kant—getting more accurate representations by inspecting, repairing, and polishing the mirror, so to speak—would not have made sense.[30]

In his most recent work, Rorty has extended and refined this critique of representationalism and realism: of the former's assumption that " 'making true' and 'representing' are reciprocal relations: the nonlinguistic item which makes S true is the one represented by S"; and of the latter's "idea that inquiry is a matter of finding out the nature of something which lies outside the web of beliefs and desires," in which "the object of inquiry—what lies outside the organism—has a context of its own, a context which is privileged by virtue of being the object's rather than the inquirer's" (*ORT* 4, 96). Instead, Rorty argues, we should reduce this desire for objectivity to a search for "solidarity" and embrace a philosophical holism of the sort found in Dewey, Wittgenstein, and Heidegger, which holds that "words take their meanings from other words rather than by virtue of their representative character" and their "transparency to the real" (*PMN* 368). Hence, Rorty suggests that we abandon the representationalist position and its privileged ocular figures and agree instead that "our only usable notion of 'objectivity' is 'agreement' rather than mirroring" (*PMN* 191).

Rorty's Deweyan reduction of "objectivity" to "solidarity" aims to dispose neatly of all sorts of traditional philosophical problems (or, as Rorty calls them, "pseudo-problems"): the problem of "skepticism" (because knowledge of things "as they are" is declared out of bounds, owing to the fact, as the later Wittgenstein realized, that "questions which we should have to climb out of our own minds to answer should not be asked" [*ORT* 7]); the relationship between (linguistic) "meaning" and (philosophical) "truth" (since to give a theory of the former is, in the absence of representationalist criteria, to give perforce a theory of the latter); the distinction between "fundamental" and "accidental" properties (since this distinction is now understood to be thoroughly rhetorical and code-bound, determined by the rules of the particular discourse in use); and, most significantly, the problem of "relativism."

  As for this last, it is worth dwelling upon for a moment because, as I have already suggested, the charge of relativism (or, as it is sometimes framed, "self-refutation") is a familiar refrain of realist critics who take issue with Rorty's abandonment of representationalism and with the broader "postmodernism" and "social constructionism" it betokens. The realist charge usually takes the following form: "How does one rule out categorical theories in principle without getting categorical? How does one universalize about theory's inability to universalize?"[31] This epistemological objection often leads, in turn, to the sort of sweeping *political* deduction we find exemplified by the Marxian theorist Norman Geras: "If there is no truth, there is no injustice. Stated less simplistically, if truth is wholly relativized or internalized to particular discourses or language games or social practices, there is no injustice.... Morally and politically, therefore, anything goes."[32] This reading of Rorty, and of postmodern "relativism" in general, has been convincingly refuted, to my mind, by thinkers such as Barbara Herrnstein Smith, Ernesto Laclau, and Chantal Mouffe, and in a different register (as we shall see) by Niklas Luhmann.[33] But no one's response is more lucid, I think, than that of Rorty himself, which is worth quoting at length:

> it is not clear why "relativist" should be thought an appropriate term... [f]or the pragmatist is not holding a positive theory of truth which says that something is relative to something else. He is, instead, making the purely *negative* point that we should drop the traditional distinction between knowledge and opinion, construed as the distinction between truth as correspondence to reality and truth as a commendatory term for well-justified beliefs.... [W]hen the pragmatist says that there is nothing to be said about truth save that each of us will commend as true those beliefs which he or she finds good to believe, the realist is inclined to interpret this as one more positive theory about the nature of truth: a theory according to which truth is simply the contemporary opinion of a chosen individual or group. Such a theory would, of course, be self-refuting. But the pragmatist does not have a

theory of truth, much less a relativistic one. As a partisan of solidarity, his account of the value of cooperative human inquiry has only an ethical base, not an epistemological or metaphysical one. (*ORT* 23–24)[34]

Rorty's Deweyan reduction of objectivity to solidarity provides the ethical basis for the pragmatist's Wittgensteinian epistemology, which insists that "it is contexts all the way down," that "we can only inquire after things under description," and that "'grasping the thing itself' is not something that precedes contextualization, but is at best a *focus imaginarius*" (*ORT* 100).

It would appear, then, that for Rorty, as for Michaels, the "outside" of belief or description (what used to be called the "referent") is always already *inside*, insofar as meaning (to borrow once again Michaels's formulation) is not filtered through what we believe, but is rather constituted by what we believe. The problem with *this* position, however, is that it immediately raises the suspicion, as Rorty recognizes, that "antirepresentationalism is simply transcendental idealism in linguistic disguise . . . one more version of the Kantian attempt to derive the object's determinacy and structure from that of the subject" (*ORT* 4). Critics of antirepresentationalism imagine "some mighty immaterial force called 'mind' or 'language' or 'social practice' . . . which shapes facts out of indeterminate goo"; and so, Rorty continues:

> The problem for antirepresentationalists is to find a way of putting their point which carries no such suggestion. Antirepresentationalists need to insist that "determinacy" is not what is in question—that neither does thought determine reality nor, in the sense intended by the realist, does reality determine thought. More precisely, it is no truer that "atoms are what they are because we use 'atom' as we do" than that "we use 'atom' as we do because atoms are as they are." *Both* of these claims, the antirepresentationalist says, are entirely empty. (*ORT* 5)[35]

But even if we agree with Rorty that "determinacy" is not exactly the issue here, the question of how the outside of belief can be accounted for *at all* certainly is. What *is* the philosophical status, exactly, of those atoms (or "atoms"?) whirling beyond the deterministic ken of our descriptions? Discussing the work of Sellars and Davidson, Rorty writes that "what shows us that life is not just a dream, that our beliefs are in touch with reality, is the *causal*, non-intentional, non-representational, links between us and the rest of the universe" (*ORT* 159). The pragmatist "believes, as strongly as does any realist, that there are objects which are *causally* independent of human beliefs and desires" (*ORT* 101); she "recognizes relations of *justification* holding between beliefs and desires, and relations of *causation* holding between those beliefs and desires and other items in the universe, but no relations of *representation*" (*ORT* 97).[36] Pragmatists do indeed accept "the brute, inhuman, causal stubbornness of the gold or the text. But they think this should not be confused with, so to speak, an *intentional* stubbornness, an insistence on being *described in a certain way*, its *own* way" (*ORT* 83).

What Rorty's response to the realist challenge here points to is a crucial realization about the relationship between "belief" and the outside of belief (what is often called "the real world"), one that enables him to avoid the double bind of immanence/transcendence that plagues the Michaels/Fish account. In the passage I am about to quote, it initially looks as if Rorty might travel down the same path (as his endorsements of Michaels and Fish in "Texts and Lumps" might lead us to expect). Rorty imagines the recalcitrant realist responding that the pragmatist, given her account, cannot "find out anything about objects at all," that "you never get outside your own head." Rorty replies, in one of the more disarming moments in the book, that "what I have been saying amounts to accepting this gambit." But, he hastens to add, one of the most central beliefs held by the pragmatist is that "lots of objects she does not control are continually causing her to have new and surprising beliefs." Hence, "She is no more free from pressure from the outside, no more tempted to be 'arbitrary,' than anyone else" (*ORT* 101).

In contrast to what we might call the "hard" version of belief propounded by the Michaels/Fish line—which holds that once you have a belief you will inhabit it "without reservation," with "no distance"[37]—Rorty here provides a "soft" account of belief, one in which beliefs are *always* held with reservations because they are held in a world in which (to use William James's picturesque phrase) our experience "has ways of *boiling over*, and making us correct our present formulas."[38] Rorty's pragmatist and Jamesian point is that there is nothing to stop you, on purely *epistemological* grounds, from believing whatever you like, but that belief itself will have consequences because it is subject to "pressure from the outside." This is the sense, I think, of Donald Davidson's assertion, which Rorty quotes approvingly, that "most of our beliefs are true"—because we are still around to talk about them! (*ORT* 9–10).

Rorty's specifically pragmatist intervention here, then, is that the imperative to theory, to reflection on belief, derives not from an essentialist "appetite of the mind" (to use James's phrase), nor from a desire for transcendence in either its realist or idealist incarnation (as the Knapp and Michaels critique of theory would have it), but rather from the strategic, adaptive, *pragmatic* value of theory that any act of intellection will ignore only at its own peril. One might well insist, with James, that the desire to theorize is "characteristically human," but "this would be like saying," Rorty writes, "that the desire to use an opposable thumb remains characteristically human. We have little choice but to use that thumb, and little choice but to employ our ability to recontextualize" (*ORT* 110). Thus, the pragmatist "takes off from Darwin rather than from Descartes, from beliefs as adaptations to the environment rather than as quasi-pictures" (which is where we set out in Fish and Knapp/Michaels) (*ORT* 10); he thinks "of linguistic behavior as tool-using, of language as a way of grabbing hold of causal forces and making them do what we want, altering ourselves and our environment to suit our aspirations" (*ORT* 81). In this way, pragmatism "switches attention from 'the demands of the object' to the demands of the purpose which a particular inquiry is supposed to serve." "The effect," Rorty concludes,

> is to modulate philosophical debate from a methodologico-ontological key into an ethico-political key. For now one is debating what purposes are worth bothering to fulfill, which are more worthwhile than others, rather than which purposes the nature of humanity or of reality obliges us to have. For antiessentialists, all possible purposes compete with one another on equal terms, since none are more "essentially human" than others. (*ORT* 110)

But here, at precisely this juncture, the pluralist imperative of Rorty's pragmatist commitment to contingency begins to break down—or, more specifically, begins to be recontained by a more complacent and uncritical sort of pluralism; for it may be true, as Rorty puts it, that "*holism takes the curse off naturalism*" (*ORT* 109), but no sooner does it resituate the *philosophical* problems of naturalism in an "ethico-political key" than it creates enormous *political* problems by reinscribing Rorty's project within the horizon of a debilitating liberal humanism and, beyond that, ethnocentrism.

Rorty's description and defense of ethnocentrism in his essay "Solidarity or Objectivity?" begins by sounding commonsensical enough: "to say that we must work by our own lights, that we must be ethnocentric, is merely to say that beliefs suggested by another culture must be tested by trying to weave them together with beliefs we already have" (*ORT* 26). But the issue that remains submerged here—and that remained submerged in the lengthy passage quoted earlier—is this: Just who *is* this generic "we" in Rorty's discourse? The problem works its way to the surface later in the same essay, where Rorty writes, again in a seemingly commonsensical moment:

> The pragmatists' justification of toleration, free inquiry, and the quest for undistorted communication can only take the form of a comparison between societies which exemplify these habits and those which do not, leading up to the suggestion that nobody who has experienced both would prefer the latter.... Such justification is not by reference to a criterion, but by reference to various detailed practical advantages. (*ORT* 29)

Even if we leave aside the idealist gesture toward "undistorted communication" (a gesture that Rorty himself has rightly criticized in Habermas),[39] and even if we subscribe to the bourgeois liberal values that Rorty inventories, the question that never gets asked here is whether all members of Rorty's society experience these "detailed practical advantages" in the way that Rorty imagines. These liberal values and freedoms may extend to all in the abstract—that is, in theory—but do they in *practice*? Clearly, the answer is no. This need not lead us to reject out of hand the liberal values Rorty regularly invokes; it is simply to point out that when Rorty claims that "we" should encourage the "end of ideology" (*ORT* 184), that "anti-ideological liberalism is, in my view, the most valuable tradition of American intellectual life" (*ORT* 64), he is staging a claim that is itself ideological through and through. What Rorty does not recognize, in other words, is that there is a fundamental contradiction between his putative desire to extend liberal advantages to an ever larger community, and the fact that those advantages are possible for some only because they are purchased at the expense of others. As Nancy Fraser puts it, the problem with "the communitarian comfort of a single 'we'" is that

> Rorty homogenizes social space, assuming tendentiously that there are no deep social cleavages capable of generating conflicting solidarities and opposing "we's." It follows from this assumed absence of fundamental social antagonisms that politics is a matter of everyone pulling together to solve a common set of problems. Thus, social engineering can replace political struggle. Disconnected tinkerings with a succession of allegedly discrete social problems can replace transformation of the basic institutional structure.[40]

In this light, it is deeply symptomatic that Rorty relies on the language of "democracy" and "community," whose homogenizing connotations mask and submerge the unevenness of power and resources in the social and economic sphere that a very different language—the language of "capital" and "class"—would force to light. The problem, as Chantal Mouffe has argued, is Rorty's

> identification of the political project of modernity with a vague concept of "liberalism" which includes both capitalism and democracy.... If one fails

to draw a distinction between democracy and liberalism, between political and economic liberalism; if, as Rorty does, one conflates all these notions under the term *liberalism;* then one is driven, under the pretext of defending modernity, to a pure and simple apology for the "institutions and practices of the rich North Atlantic democracies."[41]

From another vantage, equally disturbing in Rorty's ethnocentrism is the narrowness of the liberal *ethnos* itself. If "we heirs of the Enlightenment think of enemies of liberal democracy like Nietzsche or Loyola as, to use Rawls's word, 'mad'" (*ORT* 187), then, he acknowledges, "suddenly we liberal democrats are faced with a dilemma," for "to refuse to argue about what human beings should be like seems to show a contempt for the spirit of accommodation and tolerance, which is essential to democracy." Rorty attempts to dispense with this dilemma by insisting that "accommodation and tolerance must stop short of a willingness to work within any vocabulary that one's interlocutor wishes to use" (*ORT* 190). Again, this seems reasonable enough, but the problem is that those who are declared beyond the pale of reason during the course of Rorty's recent work include not only Nietzsche and Loyola, but also Gilles Deleuze, Jean-François Lyotard, Michel Foucault, and all those whom Rorty calls, in a *New York Times* Op-Ed piece, the "unpatriotic left" of the American academy, which "refuses to rejoice in the country it inhabits" and "repudiates the idea of a national identity, and the emotion of national pride"[42] — all those who have experienced "an apparent loss of faith in liberal democracy" (*ORT* 220). Rorty constantly invokes the liberal intellectual's dedication to expanding the range of democratic privileges, freedoms, and values, but what becomes clear in his recent work is that such an expansion can take place only *after* the democratic *ethnos* has been purified of the sort of dissent it needs to encourage.

In a position developed in detail in *Contingency, Irony, and Solidarity*, Rorty argues that we *may*, in fact, pursue the sorts of radical redescription and reinvention imagined by Deleuze, Foucault, and others; we may indeed, as Rorty writes in an essay on Foucault, pursue "inhuman thoughts" and assume the pursuits of a "knight of autonomy" — but *only* in the private realm (*EHO* 193, 194). We may "dream up as many new contexts as possible" (*ORT* 110), but "it is only when a Romantic intellectual begins to want his private self to serve as a model for other human beings that his politics tends to become antiliberal" (*EHO* 194). Rorty argues that we should avoid "the temptation to try to find a public, political counterpart" to this "desire for autonomy" (*EHO* 196). But, as Nancy Fraser deftly observes, the problem with Rorty's "partition position," as she calls it, is twofold: first, "the social movements of the last hundred or so years have taught us to see the power-laden, and therefore political, character of interactions that classical liberalism considered private" (as in feminism's well-known shibboleth "the personal *is* the political"); and second, the price of this liberal ideological containment of difference and pluralism is that "radical theorizing assumes individualistic connotations, becoming the very antithesis of collective action and political practice.... It becomes aestheticized, narcissized, and bourgeoisified, a preserve where strivings for transcendence are quarantined, rendered safe because rendered sterile" (103).

Rorty's charge against the sort of philosophy undertaken by Deleuze and the French inheritors of Nietzsche may be right: that in the interest of a philosophy of "authenticity" it too quickly and one-sidedly throws overboard (to use Vincent Descombes's phrase) "everything in which the ordinary person believes," and in so doing commits itself, as Rorty puts it, "to fantasize rather than converse," and engages in a form of thought that, insofar as it has any politics at all, is "anarchist rather than liberal."[43] But what is even clearer is that Rortyan pragmatism, as Cornel West puts it, "though pregnant with possibilities...refuses to give birth to the offspring it conceives. Rorty leads philosophy to the complex world of politics and culture, but confines his engagement to transformation in the academy and to apologetics for the modern West" (206–7). In the end, then, representationalism is undone on the *philosophical* level in Rorty's pragmatism, but only to reemerge in more powerful and insidious form on the plane of the *political*.[44]

What has not been sufficiently argued by Rorty's critics is that Rorty finds himself in this position because of his "evasion" of theory and epistemology-centered philosophy—an evasion that prevents him from exploring how the necessity of *other* beliefs, observations, or points of view *outside* of the *ethnos* in question might be generated by confronting, with a renewed *commitment* to theory, the contingency of his own. Rortyan pragmatism, in other words, expresses a *desire* for alterity but is unable to provide an adequate theory of that alterity's necessity. Part of the problem is that Rortyan pragmatism is fraught by a conflict that reaches all the way back through James to Emerson: on the one hand, it wants to be open to the outside of contingency, materiality, and social construction, but on the other it is engaged in what Tom Cohen has characterized as a series of "regressive attempts to shore up an iconic humanism, a theology of the self, a space of interiority."[45] In these terms, Rorty's divided posture toward Nietzsche and his postmodern inheritors takes its place in a long line of liberal "American exceptionalist" positions unnerved by their own flirtations with the deconstruction of liberal humanism. Such maneuvers attempt to cope with the fact, as Cohen puts it, that " 'pragmatism' *already* has two branches, two variant logics" represented by Rorty's two Nietzsches, one, "the *humanist*, the American . . . and the other, truly *other*, that of the continent, of 'theory'" (93). The problem with Rortyan pragmatism, Cohen reminds us, is "that the very discourse that advertises a turn toward a more radical materiality or *pragma* . . . ends by doing the opposite," and so indulges "an essential error in viewing the political as the equivalent of an ideology of representation or *mimesis*. One might," he continues, "ask where, instead, a more pragmatic pragmatism that is at once American *and* 'theoretical,' may see intervention as a matter of changing our very modes of *mimesis* themselves: a pragmatism which again sees epistemology as the very site of the political" (95).

It is in this light that I want to distinguish my own critique of American pragmatism from others on the left (Geras and Fraser, for example). My own view is that we need not agree with *either* the foundationalism and normativity of many of pragmatism's critics from the left, *or* with the "beyond ideology" defenses of pragmatism from within liberal humanism.[46] We must steer a third way, I believe, and pursue a pragmatism *on the site of theory*, one whose price is not the politically disabling repression of theory that has proved so tempting for pragmatism when its commitment to contingency becomes inconvenient; for, as Eva Knodt puts it in her overview of the Habermas/Luhmann debates, "if it can be shown that *any* attempt to ground a concept of rationality, whether one locates its ground in the constitutive powers of a transcendental subject or in a linguistically based notion of intersubjectivity [as in Rorty], is fraught with as many logical difficulties as the critique of such projects," then we can reveal the charges made by the normative left against Rorty and other postmodern "relativists" as "mere rhetoric," and move instead toward "a radical rethinking of the terms in which the postmodernism debate has been carried out over the past few decades" (80). Such a rethinking should endeavor to critique Rortyan pragmatism's liberal ethnocentrism, but *without* hanging that charge on the foundationalist's normative invocation of performative paradox, focusing instead on pragmatism's antitheoretical and American exceptionalist evasion of epistemology, which prevents it from pursuing the full, pluralist implications of its commitment to the contingency and constructedness of knowledge.[47]

To clarify: If it is true that *both* pragmatism and its foundationalist opponents (and not just the former) rest on grounds that are finally paradoxical, circular, or self-refuting, then you do not need normativity—and in fact you *cannot* use it—to mount the kinds of critiques of the Rortyan "we" that we find in Fraser and others.[48] As my reference to Chantal Mouffe earlier suggests (who shares with her coauthor Ernesto Laclau an essentially pragmatist response to the charge of relativism),[49] giving up on normativity and foundationalism does not mean, as Fraser and others think, giving up on political criticism. I will return to these issues at some length in my conclusion, but the point I wish to make here is that for *everyone*, *including* the realist critics of pragmatism, it is "pictures all the way down." And if that is so, then the theoretical question becomes whose picture has built into it the necessity of other pictures, other "we's." It is here, as I shall argue in the next chapter, that recent work in systems theory is especially valuable in helping us to renovate and reinvigorate pragmatism by tracing out in rigorous detail the theoretical problems of contingency and the social construction of knowledge that are raised, only to be prematurely abandonded, by Rorty. The case against Rorty's ethnocentric liberalism can and must be made in the absence of normative, foundational assurances—must be made, precisely, *pragmatically*—but it can only be made with more theory and not, as Rortyan pragmatism thinks, with less.

Coming to Terms: Stanley Cavell and the Ethics of Skepticism

In *Consequences of Pragmatism* (1982), Richard Rorty reveals that what most makes Stanley Cavell's brand of pragmatism distinctive is precisely what Rorty finds most problematic about it: its engagement with what Cavell calls "the truth of skepticism," which consists of confronting the fact, as Cavell puts it in a passage quoted by Rorty, "that the human creature's basis in the world as a whole, its relation to the world as such, is not that of knowing, anyway not what we think of as knowing" (quoted in Rorty, *Consequences* 176). Rorty's response to this commitment in Cavell is instructive enough and funny enough to be worth quoting: "What Cavell wants us not to miss is, to be sure, as important as he thinks it. But does he *have* to drag us back through Berkeley and Descartes to get us to see it? . . . Why 'the external world' *again*?" (177). What is submerged in these lines will come fully to the surface later in the essay, where Rorty argues that there are two senses of Cavellian skepticism: one that Rorty is happy to acknowledge is "as important as he thinks it," the other—which Cavell sees as tied directly to this first—that Rorty thinks is independent and "academic" in the worst sense. The first sort of skepticism, the profound sort, has "a wish," as Cavell puts it, "for the connection between my claims of knowledge and the objects upon which the claims are to fall to occur without my intervention, apart from my agreements. As the wish stands, it is unappeasable. In the case of knowing myself, such self-defeat would be doubly exquisite: I must disappear in order that the search for myself be successful" (quoted in Rorty, *Consequences* 187). For Cavell this "Kantian hope for an impossible kind of knowledge" (as Rorty puts it) defines something very close to the *condition humaine*, a condition "produced,"

Rorty writes, "by the Sartrean sense that only such an impossible sort of knowledge would overcome our terror at the sheer contingency of things" (182).

This is a condition with which Rorty is all too familiar; indeed, he finds it in every normativist, foundationalist, realist critic he confronts. "But I do not see," Rorty continues, zeroing in on the academic sort of skepticism touched on above, "how he [Cavell] can connect Pricean puzzles about getting from perceptions to non-perceptions with either Kantian longing or Sartrean terror" (182). Rorty's point—it is borne out by the fact that most readers will have no idea who "Price" is in that last sentence—is that Cavell goes wrong (goes, that is, "academic" and "philosophical") in thinking that "anybody *could* think that textbook 'English' epistemology is intimately connected with a sense of the contingency of everything" (184). "It is only when we drop the Lockean question," Rorty continues, "about whether the redness is 'out there' or 'in us' and get to the romantic Kantian question, 'Is there anything beyond the coherence of our judgments to which we can be faithful?,' that the student is hooked" (183).

It is indeed the "Kantian question" (or, more precisely, what Cavell has called the "crossing" of the lines of Romanticism and skepticism) that has remained for Cavell very much a live issue—indeed, *the* live issue—because the ethics of philosophical engagement are for him sustained only insofar as we do *not* turn away from the problem of skepticism in the way that Rorty has. Indeed, for Cavell the entire project of what he calls "moral perfectionism" rests upon doing justice to the simultaneous (and seemingly countervailing) imperatives of constructivism on the one hand and what he calls the persistent "truth of skepticism" on the other. As one commentator writes, Cavell's work asks us to confront this question: "what would happen to philosophy if we took the search for foundations from it and replaced it with the search for finding oneself?"[50] In this light, Cavell's version of pragmatism is even more firmly within the purview of humanism than Rorty's. But Cavell's humanism is, as we shall see, of a rather unusual, self-deconstructing sort, concerned as it is to bring to light not an unchanging human essence but rather a dynamic, "homeless" self of "transience" and "onwardness," a self that consists (or maybe "subsists") in always leaving itself behind.

This project is foreshadowed in *The Claim of Reason*, but it is taken up most powerfully—and less "academically," as Rorty would say—in Cavell's later work on Emerson, from *The Senses of Walden* (1972) through his most recent book, *Philosophical Passages: Wittgenstein, Emerson, Austin, Derrida* (1995). Cavell stakes out a crucial place for Emerson in the philosophical tradition, where he emerges as the founding (or rather "finding") figure of a distinctly American philosophy, and, more generally, stands as the topos where the broader currents of philosophical Romanticism and skepticism cross with unparalleled force and charge, in ways instructive for contemporary philosophy and extrapolated by the later work of Heidegger and Wittgenstein.

We can grasp more fully the importance of Cavell's reading of Emerson by briefly situating it in immediate critical context. The central interest, and the political promise, of Cavell's Emerson is that he offers an exemplary attempt to think through—but also to own, own up to—the *necessity* of selfhood without specifying, in a reductive or absolutist way, the *contents* of that selfhood. In doing so, Cavell would seem to agree with deconstruction that unreconstructed concepts of the subject of Marxian or feminist stripe are unacceptably totalizing in their reduction of the full complexity of the subject in the name of class or gender.[51] At the same time, however, Cavell's *interpretation* of this fact differs from deconstruction's, sometimes pointedly. Cavell is quite clear on this in several places, most of all in his postscript "Skepticism and a Word concerning Deconstruction" (in *In Quest of the Ordinary*), where he agrees with deconstruction that "language is inherited, learned, always already there for every human," but finds the ethical inference drawn from this assumption by the Emerson/Wittgenstein line different from what we typically find in deconstruction. In Cavell's reading, deconstruction will typically take that assumption to emphasize that the distinction between "quoted words and their originals" (a figure we will return to in a moment with Emerson's transvaluation of Descartes) is empty, whereas the Emerson/Wittgenstein line "will see that emphasis as deflecting attention, as rushing too quickly away, from the act or encounter entailed in the historical and individual process of inheriting,"[52] a process that is the very site of what Cavell's most recent work calls the ethical assumption or "arrogation" of the voice.[53]

It is worth lingering over this difference with deconstruction for a moment, because it is here that the *pragmatism* of philosophy for Cavell is at stake. For Cavell, the aim of philosophy is "not to undermine but to underline such distinctions as that between quoting and saying.... Then style and its obligations become the issue—what I might call the address of language, or the assumption of it, perhaps the stake in it" (*Quest* 133). For Cavell, this is the pragmatic point that the Derridean wholesale critique of voice misses, the possibility that there can be not only a voice of metaphysics, but also a voice of what Cavell calls the "ordinary." In a recent interview, Cavell finds Derrida's critique of Austin in flight from the ordinary even as Derrida himself acknowledges "that there are 'effects' of the ordinary. . . . It matches the moment in which," Cavell continues, "the academic skeptic says *of course* we know that there are tables and pens and hats, etc., *for practical purposes.* But what are these practical purposes?" (*PP* 74). "That skeptic," Cavell continues,

> says, or takes the tone in which, he cannot shake the knowledge he has discovered in his closet, namely that we can never know with certainty that there are things and other minds, etc., while at the same time he recognizes that when he leaves his isolation, comes out into the company of others, plays backgammon with his friends, that he will forget this terrible truth — as, of course, he knows he should, being a sociable creature who does not crave insanity. Now this gesture, this "for practical purposes I know," . . . further[s] the air of implication that there is a something more to do — a further reality to assess, a fullness of certainty to apply — than human beings can compass. (Derrida also denies certain understandings of this "more." Is this my contradiction, or his?) Of course — so runs the air — I know your signing the check means that it is your signature, not mine; but only for practical purposes; this is no assurance of my tie to a metaphysically independent world. (74)

Hence "the metaphysician in each of us," Cavell concludes — either in positive form (with the foundationalists of the philosophical tradition) or in negative form (with deconstruction) — "will use metaphysics to get out of the moral of the ordinary, out of our ordinary moral obligations" (74–75). In this way, deconstruction for Cavell refuses "the responsibility you bear — or take, or find, or disclaim — for your words" (*Quest* 135).[54] It too often only "theatricalizes the threat, or the truth, of skepticism," because it "names our wish (and the possibility of our wishing) to strip ourselves of the responsibility we have in meaning." And as such, deconstruction constitutes not a new way of thinking about ethics or politics, but a refusal of both. "Such courses," Cavell writes, "seem to give up the game; they do not achieve what freedom, what useful ideal of myself, there may be for me, but seem as self-imposed as the grandest philosophy — or, as Heidegger might almost have put it, as unself-imposed" (*Quest* 131).[55]

My primary concern here is not Cavell's differences with deconstruction, but those differences do help to underscore the stakes in Cavell's reading of Emerson. For Cavell, the power and promise of Emersonian selfhood begins with the essential rigor of its confrontation with the truth of skepticism, its dedication, after *Nature*, to understanding the fullest implications of skepticism's unhappy fact: "that the world exists as it were for its own reasons" (*This New* 79).

In the philosophical tradition, the classic attempt to reach a "settlement" with this fact of skepticism is the Kantian one in the *Critique of Pure Reason*, whose problems Cavell sums up in *In Quest of the Ordinary*:

> The dissatisfaction with such a settlement as Kant's is relatively easy to state. To settle with skepticism...to assure us that we do know the existence of the world, or rather, that what we understand as knowledge is *of* the world, the price Kant asks us to pay is to cede any claim to know the thing in itself, to grant that human knowledge is not of things as they are in themselves. You don't—do you?—have to be a romantic to feel sometimes about that settlement: Thanks for nothing. (31)[56]

According to Cavell, Emerson after *Nature* does not work around Kant but rather works through him; he takes the claims of skepticism more seriously than Kant himself did by turning the Kantian position back upon itself and subjecting the very terms of Kant's argument to transcendental deduction. What Kant conceived as a problem of thinking and philosophy, Emerson will confront more rigorously as a problem of language and writing as well, so that the "stipulations or terms under which we can say anything at all to one another" will themselves be subjected to scrutiny (*This New* 81).

The beguiling and specific value of Emerson's work, according to Cavell, thus derives from a twofold recognition. First, Emerson realizes, in the wake of skepticism, that "philosophy has to do with the perplexed capacity to mourn the passing of the world" (*This New* 84); but second, and more important, this means not that our language is now philosophically impotent, but that the "terms" we strike with existence, the "terms" we use to write that agreement, are in a very real sense all we have. Emerson characterizes it this way in the late essay "Fate": "Intellect annuls fate. So far as a man thinks, he is free." "This apparently genteel thought," Cavell writes, "now turns out to mean that...our antagonism to fate, to which we are fated, and in which our freedom resides, is as a struggle with the language we emit, of our character with itself" (*Quest* 40).[57]

This will be clearer, perhaps, in one of the more important moments in Cavell's most important meditation on the meaning of Emerson, the essay "Finding as Founding: Taking Steps in Emerson's 'Experience,'" which stands at the center of *This New Yet Unapproachable America*. There, he unpacks a crucial passage from Emerson's essay that also serves as the epigraph for the 1988 Carus Lectures collected in *Conditions Handsome and Unhandsome*: "I take this evanescence and lubricity of all objects," Emerson writes, "which lets them slip through our fingers then when we clutch hardest, to be the most unhandsome part of our condition." According to Cavell, "the unhandsome" in this passage names not so much the fact that we cannot finally know, apprehend, or contact the world of things, but "rather what happens when we seek to deny the stand-offishness of objects by clutching at them; which is to say, when we conceive thinking, say the application of concepts in judgments, as grasping something, say synthesizing" (*This New* 86). Conversely, the "most handsome part of our condition," "clutching's opposite," is the realization that "the demand for unity in our judgments, that our deployment of concepts, is not the expression of the conditionedness or limitations of our humanness but of the human effort to escape our humanness" (86–87).

In "Experience," Emerson will surrender that escapist project with the essay's opening question — "Where do we find ourselves?" — and thus move toward what Cavell calls the "overcoming of thinking as clutching." And that redefinition of the task of philosophy, in turn, will lead Emerson into an investigation of the "conditionedness" and limitations of our terms of existence. But before any of this can take place, Emerson must work through what Cavell calls a necessary "mourning" for the passing of the world, which in "Experience" is mediated by Emerson's grieving over the impotence of his grief at the death of his son. Unlike deconstruction, Emerson does not overleap this task of mourning but rather confronts it, and it is that mourning which constitutes not a "deflection" of skepticism but rather a "respect for it, as for a worthy other" (88). If, in Kant's settlement with skepticism, "reason proves its power to itself, over itself" by discovering the difference between the mere appearances of which it can have knowledge and the Kantian *Ding an sich* of which it cannot (*Quest* 30), then Emerson's mourning of the "unhandsomeness" of the world entails, in Cavell's words, the "recognition not of the uncertainty or failure of our knowledge but of our disappointment with its success" (*This New* 88).

This continuation of the task of philosophy after philosophy is, in a sense, impossible constitutes "our poverty," Cavell tells us in *The Senses of Walden*, "what hope consists in, all there is to hope for."[58] The philosophical self is thus condemned not to "founding" its existence by reference to traditional philosophical categories, but rather to what Cavell's reading of "Experience" calls "finding." And likewise, the project of the self becomes not so much "dwelling," the "inhabitation and settlement" of the world we find in the late Heidegger, but rather moving through it, "lasting" by journeying, what Cavell in "Thinking of Emerson" calls "the task of onwardness" (*Senses* 136–38).

Perhaps Cavell's clearest and most compelling account of this "transient" philosophical self — Emerson as vagrant, "The philosopher as the hobo of thought" (*This New* 116)—takes place in his essay "Being Odd, Getting Even (Descartes, Emerson, Poe)." In what Cavell playfully calls "the story of the discovery of the individual," Emerson's "Self-Reliance" constitutes an important revision, in the light of skepticism, of the Cartesian cogito. The central fact of what Cavell calls the Cartesian "proof" of selfhood is the "discovery that my existence requires, hence permits, proof (you might say authentication)—more particularly, requires that if I am to exist I must name my existence, acknowledge it" (*Quest* 106). From this vantage, Emerson's allusion to Descartes in "Self-Reliance" assumes new significance: "Man is timid and apologetic," Emerson writes; "he is no longer upright; he dares not say 'I think,' 'I am,' but quotes some saint or sage." For Cavell, the power and rigor of Emerson's revision of the Cartesian proof of selfhood is that it "goes the whole way with Descartes's insight" by continuing to require the proof of selfhood *without* allowing us to rely on a preexistent, "quotable" content to underwrite the proof. The "beauty" of Emerson's answer to Descartes, Cavell continues,

> lies in its weakness (you may say its emptiness)—indeed, in two weaknesses. First, it does not prejudge what the I or self or mind or soul may turn out to be, but only specifies a condition that whatever it is must meet. Second, the proof only works in the moment of its giving, for what I prove is the existence only of a creature who *can* enact its existence, as exemplified in actually giving the proof, not one who at all times does in fact enact it. The transience of the existence it proves and the transience of its manner of proof seem in the spirit of the *Meditations*. . . . Only in the vanishing presence of such ideas does proof take effect—as if there were nothing to rely on but reliance itself. (*Quest* 109)

The specific genius of Emersonian self-reliance, then, lies in the fact that it is "not a state of being but a moment of change, say of becoming—a transience of being, a being of transience" (*Quest* 111). And the project of Emersonian self-reliance—or what Cavell will come to call "moral perfectionism"—is thus driven by, and follows through on, the challenge of skepticism.

But as persuasive and appealing as Cavell's reading of Emersonian individualism is, it is difficult to see how such a self could ever engage in social and political praxis—that is, in the directed transformation of the social and material conditions of freedom. In *The Senses of Walden, In Quest of the Ordinary*, and *This New Yet Unapproachable America*, when the question of praxis and politics *does* arise, Cavell usually brackets it or works through it with uncharacteristic rapidity. In *This New Yet Unapproachable America*, for instance, he notes that Emerson's question "Why not realize your world?" at the end of "Experience" "necessitates taking up philosophically the question of practice" (*This New* 95). But when Cavell *does* take it up, he will only go so far as to say that "For Emerson, as for Kant, putting the philosophical intellect into practice remains a question for philosophy" (95). For Cavell, "The first and last answers in 'Experience' to the question of realizing philosophy's worlds are recommendations to ignorance—not as an excuse but as the space, the better possibility, of our action. In the second paragraph: 'We do not know today whether we are busy or idle'" (96).

    The problem with this deferral or bracketing of the question of politics is that it does not critique or account for the disabling contradictions of Emerson's own concept of action and praxis so much as it reproduces them—all of which might be of less moment were it not for the fact that Emerson's own writings on the question of action underscore time and again the irreconcilable relationship between praxis and his brand of individualism. That much is unmistakable in essays like "Politics" or "The American Scholar," where Emerson tells us that "The world of any moment is the merest appearance. Some great decorum, some fetish of government, some ephemeral trade, or war, or man, is cried up by half mankind and cried down by the other half, as if all depended on this particular up or down."[59] It is not that material forms and institutions like books, laws, and governments are for Emerson completely worthless or impotent; it is rather that their value—their capacity to carry the truth of Reason or the power of Spirit—is finally dependent not on the concrete specificity of the particular action or social form, but rather on the individual's ethical relationship to it. For Emerson, the material character of social actions and forms renders them always epiphenomena of the subject, whose project of edification they can serve or express but never determine.[60]

    For Cavell too, the value of action is not that it has real effects on the shared, material world of others, but rather that it always returns to the self as a "resource" in an essentially isolate journey of moral perfectionism. And indeed, Cavell says as much at the end of his discussion of Emerson in *This New Yet Unapproachable America*: "Emerson," Cavell writes, "may be understood to be saying that . . . the mood of the one prepared to be useful to the world is different from that of the one prepared to adapt to it. . . . The existence of one of these worlds of life depends on our finding ourselves there" (*This New* 96). Yes, but only if the exteriority and materiality of practice, its "world," is simply a negative moment in a fundamentally private, individual drama. In which case, we are forced to say—borrowing Emerson's phrase in "Compensation"—that all actions finally *are* "indifferent."

    Contrary to its own designs, Cavell's account is most convincing where it underscores how the Emersonian project relentlessly *prevents* anything like a social praxis or politics of solidarity. In "Self-Reliance," Cavell writes, Emerson tells us in so many words that "politics ought to have provided conditions for companionship, call it fraternity; but the price of companionship has been the suppression, not the affirmation, of otherness, that is to say, of difference and sameness, call these liberty and equality. A mission of Emerson's thinking is never to let politics forget this" (*Quest* 119). This seems to me perfectly accurate in its account of the negative, critical power of Emerson's defense of difference, freedom, and all that he means when he writes "*Whim.*" But the larger point here is that Emerson's vision of freedom is so pure that it prevents political praxis and collective action in its antinomian insistence that the fluid "active soul" be true only to itself, above all compromise, beyond all cooperation. This is the Emerson who calls on us to "shun father and mother and wife and brother when my genius calls me," who insists that "When the good is near you ... you shall not discern the footprints of any other; you shall not see the face of man; you shall not hear any name."[61] What gives Emerson's position its critical force, in other words, is precisely what makes it radically *antisocial*, not an agent of praxis but a continual, insistent refusal of it. From this vantage, it is true, but not necessarily good news, that what makes Emerson a model philosopher for Cavell is that his work "asks the philosophical mood so purely, so incessantly, giving one little other intellectual amusement or eloquence or information, little other argument or narrative, and no other source of companionship or importance, either political or religious or moral, save the importance of philosophy, of thinking itself" (*Senses* 152).

To Cavell's credit, he is the first to acknowledge that when he comes to consider the "sociality" of Emerson's text, what he finds is the overwhelming presence of "unsociability or ungeniality, its power to repel, its unapproachability marked as its reproachfulness." In fact, for Cavell's Emerson, this "unsociability" is most of all what constitutes "a life's, or a text's, genius," a "power" (*This New* 12) that comes to function in the later Emerson in the services of his recognition "that with each word we utter we emit stipulations, agreements we do not know and do not want to know we have entered," as Cavell puts it, "agreements we were always in, that were in effect before our participation in them." But what looks to Cavell like Emerson's ungenial rejection of institutions of authority and conformity can be seen from another perspective as not only un- but *anti*social and, in a strict sense, reactionary. This is so because Emerson's unruliness arises in repulsion not to a certain *form* of the social, but rather to the social *as such*, to the fact that the Emersonian self is not the origin of the preexistent "agreements" and "stipulations" that constitute the shared space of the social and the other. But rather than making it possible to think a social practice that might make those preexistent "stipulations" of self-hood less alienating, Cavell's Emerson instead refuses sociality altogether—that is his seductive power and severe limitation.

In two particular texts—the essay "Politics as Opposed to What?" and the collection *Conditions Handsome and Unhandsome*—Cavell attempts to turn this "repellent" fact about Emersonian subjectivity into a virtue. How might this seeming "political liability," as he phrases it—the fact that, for the Emersonian, "the politics of philosophical interpretation" appears to rest upon "a withdrawal or rejection of politics, even of society, as such"[62]—be transformed into a sign of philosophy's labor on behalf of freedom and democracy? Under the sign of Emerson and Thoreau, the "self-containedness" of the philosophical text "should be interpretable politically," Cavell writes, "as a rebuke and confrontation" of all forms of tyranny and oppression that reside in society as we know it—a "withholding of assertion" that should be read as "the foregoing of domination" ("Politics" 199). But, as Cavell realizes, we are immediately forced to ask how this rejection of the social on behalf of freedom is to be distinguished from antidemocratic elitism. In *Conditions Handsome and Unhandsome*, Cavell's response to this dilemma takes the following form: the individual's own "ungenial" journey of moral perfectionism must be seen, he argues, as "representative" (to use Emerson's term) of the journeys of others. If "the task for each is his or her own self-transformation, then the representativeness in that life may be recognized," Cavell writes, "not only in one's own past selves but also in the selves of others." In this way, "Emerson's writing works out the conditions for my recognizing my difference from others as a function of my recognizing my difference from myself" (*Conditions* 52–53). And the Emersonian idea of "representativeness" thereby transforms what looks at first glance like an elitist perfectionism into an other-regarding social project.

But it is clear in the Introduction to *Conditions Handsome and Unhandsome* that Cavell's unabashedly liberal individualism will reproduce the troubling implications for the self's relations with others that are already familiar to us from Emerson. Cavell, like Emerson, takes for granted the existence of that very thing—call it democracy, justice, equality—whose existence it is his burden to demonstrate. In Cavell's critique, democracy is always virtual, always to be achieved—as indeed it must be, for the claim of Emersonian perfectionism to antielitism rests on our siding with "the next or future self, which means siding against my attained perfection (or conformity), sidings which require the recognition of an other" (31). At the same time, however—and this, I think, is the most fundamentally troubling part of Cavell's formulation—we are obliged to agree that democracy *already* exists, if only in imperfect form, if we are to enter *at all* what Cavell calls the "conversation of justice."

This problematic stance toward the political is especially clear in the Introduction to *Conditions Handsome and Unhandsome*:

> You cannot bring Utopia about. Nor can you hope for it. You can only enter it. (If you cannot imagine entering it, then either you think that the world you think must look very different from the world you converse with, or else you find that the world you converse with lacks good enough justice. In this way the imaginability of Utopia as modification of the present forms a criterion of the presence of good enough justice....) (20)

At first glance, Cavell in this passage seems only to name an epistemological truism by reminding us that all concepts of Utopia are constructed out of the discursive resources of the present—hence you can only "enter" Utopia, not bring it about, because it is always already there before you are, in the present and its discursive conditions of possibility. But there is more going on here, and more at stake, than at first appears. After all, what a curious—and curiously self-defeating—formulation this is: if you can imagine Utopia, then the justice of the present society is "good enough." And if it is *not* good enough, then recourse to the Utopian ideal is not available to you. In which case, we are forced to say that if Utopia must be a version of, a perfection of, presently existing society, and not a break or rupture from it, then who needs Utopia anyway? One would have thought that the necessity and power of Utopian thought was that it challenged, rather than took for granted, the assumption that we exist in a world of "good enough justice."[63]

All of these issues reach a head, it seems to me, in an essay on Emerson from Cavell's latest book, *Philosophical Passages*, titled "Emerson's Constitutional Amending: Reading 'Fate.'" There, Cavell reprises his reading of Emerson's "aversive" thinking in *Conditions Handsome and Unhandsome*—as in the famous Emersonian assertion that "self-reliance is the aversion of conformity"—and acknowledges once again that Emerson's "writing and his society incessantly recoil from, or turn away from one another" (13). But Cavell realizes that the stakes here are very high indeed because of the immediate social and political context of Emerson's essay:

> Could it be that the founder of American thinking, writing this essay in 1850, just months after the passage of the Fugitive Slave Law, whose support by Daniel Webster we know Emerson to have been unforgettably, unforgiv-

ingly horrified by, was in this essay not thinking about the American institution of slavery? I think it cannot be. Then why throughout the distressed, difficult, dense stretches of metaphysical speculation of this essay does Emerson seem mostly, even essentially, to keep silent on the subject of slavery, make nothing special of it? It is a silence that must still encourage his critics . . . to imagine that Emerson gave up on the hope of democracy. (15)

But not quite, Cavell argues, for what Emerson is after in this essay is a way of participating in the conversation of democracy that is *not* like the "crowing about liberty by slaves, as most men are" that Emerson denounces in "Fate." It is Emerson's "refusal of crowing" that is important here for Cavell, the suggestion that "there is a way of taking sides that is not crowing, a different way of having a say in this founding matter of slavery" (17). And that way, of course, is for Cavell the way of philosophy, the way, as Emerson famously puts it, of "Man Thinking": "If slavery is the negation of thought," Cavell writes, "then thinking cannot affirm itself without affirming the end of slavery. . . . Philosophy cannot abolish slavery, and it can only call for abolition to the extent, or in the way, that it can call for thinking, can provide (adopting Kant's term) the incentive to thinking" (29). In this manner, Cavell argues, "the absoluteness of the American institution of slavery, among the forms human self-enslavement," calls forth "the absoluteness of philosophy's call to react to it, recoil from it." And hence — as Cavell puts it with wonderful compression — "the apparent silence of 'Fate' might become deafening" (36).

This, it seems to me, is Cavell's most strenuous, and finally most strained, attempt to transvalue what he calls Emerson's antihumanism — his "working 'against ourselves,' against what we understand as human (under)standing" (36) — into a more profound, more radical humanism. Cavell realizes, of course, that this is perilous business, and he asks with admirable candor,

> Is philosophy, as Emerson calls for it . . . an evasion of actual justice? . . . I think sometimes of Emerson, in his isolation, throwing words into the air, as aligned with the moment at which Socrates in the *Republic* declares that the philosopher will participate only in the public affairs of the just city, even if this means that he can only participate in making — as he is now doing — a city of words. (31)

These reservations and the contradictions that generate them are symptomatic, I think, of the reductive idealism of Cavell's concepts of politics and the social, an idealism revealed by a series of important slippages throughout the attempted political recuperation of Emerson in *Conditions Handsome and Unhandsome* and *Philosophical Passages*. Through the familiar liberal pluralist metaphor of the "conversation" of justice made famous by Rorty, politics ("democracy") is narrowed down to its symbolic, communicative (as opposed to economic and material) dimension. And then *that* is further reduced, in turn, to the style or writing, the "terms," of particular philosophical ideals that maintain a last, symbolic tie to politics by virtue of their "representativeness."

"Emersonian Perfectionism," Cavell writes, "is not primarily a claim as to the right to goods (let alone the right to more goods than others) but primarily as to the claim, or the good, of freedom" (*Conditions* 26). In fact, this is a massive understatement; Cavell's reading of Emersonian Perfectionism is not a claim *at all* about the right to goods, because it emphasizes time and again just how beside the point the right to goods is in Cavell's conception of moral perfectionism and the conversation of justice—a point carried by his pun on the "good" of "freedom." Or again: "The issue of consent," Cavell tells us, "becomes the issue of whether the voice I lend in recognizing a society as mine, as speaking for me, is my voice, my own. If this is perfectionism's issue, it should indicate why perfectionist claims enter into the conversation of justice" (*Conditions* 26–27). But one need only register the point that all "voices"—even in a society with "good enough justice"—do not enter the "conversation" of justice with the same sort of power, authority, and resources to make themselves heard and binding. Such "voicing"—the democratic rationale for moral perfectionism—is enabled or compromised by goods and resources that are not equally or evenly distributed. Cavell's vision of moral perfectionism and the conversation of justice is thus blind to the real inequality of goods and the power they confer—call them the resources of voicing—in the realm of everyday material life by telling us that freedom to enter the conversation of justice and the project of democracy is shared equally by all in the realm of ideas.

Cavell's reading of "transient" Emersonian selfhood and its deconstruction of the Cartesian cogito is quite faithful, I think, to the spirit and letter of Emerson's own position, but in a way that is precisely the problem, because Cavell's attempt to articulate a politics of individualism, like Emerson's own, is undermined by its inability to escape the logic, structure, and alienating social implications of private property. In the revised edition of *The Senses of Walden*, Cavell is emphatic that the kind of subjectivity envisioned by Emerson is not one of possession, not structured by the logic of property: "This possessing is not—it is the reverse of—possessive; I have implied that in being an act of creation, it is the exercise not of power but of reception" (*Senses* 135).

It is true—particularly early in his career—that Emerson's stance toward the logic of property often appears to be critical and subversive, as in the famous moment, early in *Nature*, where he writes:

> Miller owns this field, Locke that, and Manning the woodland beyond. But none of them owns the landscape. There is a property in the horizon which no man has but he whose eye can integrate all the parts, that is, the poet. This is the best part of these men's farms, yet to this their warranty-deeds give no title.[64]

Here and elsewhere, Emerson's aim is apparently to appropriate the rhetoric of property and cagily put it to the services of his own brand of idealism, which tells us that people too often mistake social and historical institutions (like property) for what Emerson elsewhere calls simply "the reality."[65] But the more fundamental point here is that Emerson's idealism rejects the logic of property only to reinscribe it at the very heart of his critique. The problem with the deed and title of Miller and Locke, after all, is not that they are forms of ownership, but rather that they are not, like the poet's more perfect possession, ownership *enough*.

In *Nature*, Emerson had already reminded us that property "is a preceptor whose lessons cannot be foregone," that it is "the surface action of an internal machinery, like the index of the face of a clock," that our relationship to it reflects our "experience in profounder laws."[66] And this belief in the more than merely historical truth of the logic of property would only solidify in his later career, and particularly in the period 1842 to 1850, where we increasingly find what Sacvan Bercovitch has called Emerson's "unabashed endorsements . . . of what can only be called free enterprise ideology."[67] More and more, in essays like "Compensation" and "Wealth," Emerson takes to heart the mutually expressive relationship between property and selfhood, and participates in a kind of capitalization of Spirit and a spritualization of capital, consistently finding, as Michael T. Gilmore puts it, "economic categories applicable to the operations of the Soul."[68]

It is important to realize that this is true of Emerson's career both late *and* early. Emerson does not simply offer the logic of property as one of many metaphors for self-reliance, but instead insists that property is, in fact, *coterminous* with the self—where it comes, there comes self-reliant man. Emerson's apparent rejection of wealth and property only serves to make way for a more significant *perfection* of it—hence his claim that "accidental" property, unlike the "living" kind, "is not having."[69] The Emersonian self loses the world of experience and possession, but only to enable a more fundamental—and fundamentally transhistorical—

kind of possession. As Emerson characterizes it toward the end of "Experience," "We must hold hard to this poverty, however scandalous, and by more vigorous self-recoveries, after the sallies of action, *possess our axis more firmly*" (287; my emphasis).

This essential logic of Emersonian self-possession is registered with maximum force and compression at a key moment in "Experience," where Emerson writes, "All I know is reception. I am and I have: but I do not get, and when I have fancied that I have gotten anything, I found I did not" (289). Cavell reads this passage as "an explicit reversal of Kant on knowing," as a continuation of Emerson's project of coming to terms with skepticism and the "unhandsomeness" of the world. The self, Emerson tells us, cannot "get" because the material and historical world of property and circumstance cannot "touch" the self; they only leave him as they find him—as Emerson icily writes of the death of his son—"neither better nor worse." Cavell's treatment of this passage, however, skims over precisely what seems to me most crucial about it—"without pursuing this invitation," as he freely acknowledges, "to think about the structural relationships of epistemology with economy, of knowing with owning and possessing as the basis of our relation with things" (*This New* 104). But when we *do* pursue that invitation—as the pervasive rhetoric of property in the Emersonian text everywhere compels us to—we find that, here again, Emerson does not so much question the constitution of the self by the logic of property, the construction of "am" by "have," but rather confirms it. You do not "get" in the world of history and experience because you always *already* "have" everything you can have, and that is, of course, the very private property of *self*. By these lights, Cavell's reading of Emerson's valuation of "whim" in "Self-Reliance" makes perfect sense, though it seems at odds with Cavell's own critical intention: it is "something which is of the least importance, *which has no importance but for the fact that it is mine*, that it has occurred to me, becomes by that fact alone of the last importance; it constitutes my fate" (*Senses* 154; my emphasis).

My aim in all of this is simply to take Cavell at his own word—and then some; that is, Emersonian "whim" may seem only an emptiness, but Cavell, better than anyone, reminds us time and again that the forms, the "terms" that structure that emptiness—under which nothing stands, as it were—are in a very real sense all we have. But those terms themselves, it needs to be added, come to the philosophical text *already* worked and structured by a broader social context, and Cavell's formulation needs to take account of the fact that in that context—in *our* context—the logic and structure of property is not simply one among many, but is rather the central fact of social life under capitalism. Or, to put it another way, Emerson's use of the logic of property reproduces not only a discursive structure—

within which the relation of property to other logics is relatively equal or symmet-rical—but an economic one as well, within which nothing could be farther from the truth. This much more sharply asymmetrical relationship to the economic struc-ture of property, which may be reinforced by *discursive* structures such as Emer-son's, is a product of the simple fact of scarcity. As Perry Anderson reminds us, "ut-terance has no *material* constraint whatever: words are free, in the double sense of the term. They cost nothing to produce, and can be multiplied and manipulated at will, within the laws of meaning. All other major social practices are subject to the laws of natural *scarcity:* persons, goods or powers cannot be generated *ad libitum* and *ad infinitum.*"[70]

That fact, too, we need to remember, is part of the *full* "condi-tionedness" of the "terms" by which subjectivity is constructed, negotiated, and en-forced. "Freedom is obeying the law we give to ourselves," Cavell writes; "which is to say: freedom is autonomy" (*Conditions* 28). But in Cavell, as in Emerson, the con-ception of freedom as pure autonomy—and autonomy, in turn, as the attribute of a self modeled on private property—operates as what we could call, to borrow again Jameson's phrase, an ideological "strategy of containment." Here, a Utopian desire for freedom and equality is expressed by a logic (in this case individualism as self-possession) that prevents that freedom from ever being anything other than imagi-nary—which renders it, in a word, self-contradictory.[71] If selfhood is conceived in terms of self-possession of one's own person, capacities, and energies, then the self's freedom, as C. B. McPherson puts it in his classic account, consists in its "right to enjoy them and use them and to exclude others from them; what is more, it is this property, this exclusion of others, that makes a man human."[72] As Marx character-izes it, this type of individual will see in others "not the *realization* but the limitation of his own freedom," because freedom for such a self means the right "to enjoy and dispose of one's resources as one wills."[73] All of which seems to be borne out by the central fact about individual freedom in Emerson and in Cavell: it is above all free-dom to be alone, adrift in the vacuum of autonomy. As for that kind of self, Emer-son by mid-career had already penned his lament: "God delights to isolate us every day" ("Experience" 280).

How you feel about that isolation, and how you feel about Cavell's use of Emerson, will depend in large part on what you think philosophy is—or rather, to put a somewhat finer point on it, what you think it can afford to be. As I have already indicated, my response to Cavell's attempt to transvalue Emerson's "Fate" by suggesting that philosophy can fulfill its social function only by "turning away" from society, that it can condemn slavery only by keeping silent about it and in that silence exercise its own socially exemplary freedom—my response to that view of philosophy is "so much the worse for philosophy." That does not mean that what Cavell regards as the function of philosophy is not a very important *part* of philosophy: its imperative that we never cease to think the fact of contingency, difference, and openness, that philosophy never allow politics to forget that. But that is only a part of philosophy.

To put it as briefly as possible, philosophy can have, as Cavell wishes, its autonomy and necessary insulation from the polemics and public discourses that are "an evasion, or renunciation, of philosophy" (*PP* 31)—from all those things that Emerson says in "Fate" are "in the air." *Or* it can have its privileged, socially representative function. But it cannot have both. Philosophy may indeed be viewed, as Cavell characterizes it, as "the achievement of the unpolemical, of the refusal to take sides" (*Pitch* 22), of "finding your neutrality," which is the only way of "becoming what you are" (35). But if that is the case, then philosophy by definition declares as *not its problem* a whole host of material, discursive, and institutional challenges that bear directly on the creation of the material conditions whereby philosophy's ideal of freedom, which it rightly takes up, might become a reality, *precisely* in the world of the "ordinary," the "everyday." But if philosophy so declares itself, then it also ceases to be socially representative, because it then becomes only one of many specialized discourses and practices (including, of course, polemics) at work in the social field on behalf of freedom and difference, discourses that likewise declare other things (like philosophy) not *their* problem. That recognition would in effect take for granted the postmodern positions of Lyotard, Luhmann, and others, which assert the "horizontal" (or nonhierarchical) functional differentiation of social systems and the irreducible differences between discourses and language games, a position that Cavell is unwilling to take.

Of course, philosophy *could* admit that the problems I have touched on *are* its problems as well, could grant that the achievement of "neutrality" and insularity from other social discourses—the imperative of pure theory—is only one of its functions. That position, it seems to me, would constitute a more *modern* than postmodern solution to the problem, one whose most notable embodiment probably remains the middle to later Sartre. Philosophy would then indeed be a socially representative, privileged discourse—but it would cease to be philosophy in Cavell's sense. Of course, Cavell would be the first to point out that such a decision about how and when to let philosophy "come to an end" is itself a philosophical question. My point is simply that it is not *only* a philosophical question, and sometimes, in some contexts, maybe not even *mainly* a philosophical question. So it is that Cavell's skepticism promises to meet head-on the problem of theory's "outside," but does so only to reinscribe that outside within an idealized and isolated sphere called "philosophy." And in doing so, Cavell truncates the very pragmatism his commitment to contingency promises. For Cavell as for Emerson, the question that opens "Fate" — "How shall I live?" — turns out to be a philosophical question only.

T W O

Systems Theory

Maturana and Varela with Luhmann

IN THE current social and critical moment, perhaps no project is more overdue than the articulation of a posthumanist theoretical framework for a politics and ethics not grounded in the Enlightenment ideal of "Man." Within postmodern theory, that humanist ideal is critiqued as forcefully as anywhere in the early and middle phase of Michel Foucault's career, whose "genealogical" aim is to "account for the constitution of knowledges, discourses, domains of objects, etc., without having to make reference to a subject which is either transcendental in relation to the field of events or runs in empty sameness throughout the course of history" by virtue of his—and it must be "his"—privileged relation to *either* the presence or the absence of the phallus, language, the symbolic, property, productive capacity, toolmaking, reason, or a soul.[1] In Foucault, however, this call for posthumanist critique is more often than not accompanied, as many critics have noted, by a dystopianism that imagines that the end of the humanist subject is the beginning of the total saturation of the social field by power, domination, and oppression.[2] And the later Foucault, as if compensating for his early dystopianism, evinces a kind of nostalgia for the Enlightenment humanism powerfully critiqued in his early and middle work but approached much more sympathetically in the *History of Sexuality* project.[3]

But posthumanist theory need not indulge either Foucauldian dystopianism or its compensatory nostalgia for the subject to critique the ethical and political separation of the human from the nonhuman on the basis of what Bruno Latour has recently called all the "magnificent features that the moderns have been able to depict and preserve": "the free agent, the citizen builder of the Leviathan, the distressing visage of the human person, the other of a relationship, consciousness, the *cogito*, the hermeneut, the inner self, the thee and thou of dialogue, presence to oneself, intersubjectivity." As Latour recognizes, posthumanist theory cannot proceed simply by historicizing the human; instead, he argues, "we first have to relocate the human, to which humanism does not render sufficient justice."[4] And in this project of relocation, historical and dialectical means of resituating the human are not enough.

Indeed, one need only think of the difficulties experienced by the Marxist tradition in theorizing the problem of ecology to see that the limitations of humanism and the legacy of the Enlightenment episteme are not solved by dialectical historicization alone.[5] Even within the Marxist tradition, a number of theorists have recognized that Marxism's liberation of "the total life of the individual" (to borrow Marx's phrase from *The German Ideology*) is purchased at the expense of its brutal objectification of nature and the nonhuman — a dynamic deeply symptomatic, in turn, of its Enlightenment inheritance that imagines that man-the-producer liberates himself insofar as he fully exploits and raises himself above that object and resource called "nature." No one in the Marxist tradition was more keenly aware of this than Theodor Adorno, whose "negative dialectics" may be viewed (as critics as diverse as Fredric Jameson and Drucilla Cornell have suggested) as a kind of limit case in the attempt to "relocate" the human by historical, dialectical means. Adorno lamented that the historical dialectic of traditional Marxism would turn the whole of nature into "a giant workhouse" for an essentially imperialist subject, and he proposed instead "a thoroughgoing critique of identity" and of "the Concept" that might enable thought to relocate the human in a field characterized by what he called "the preponderance of the object" — the preponderance, that is, of the heterogeneity, multiplicity, and nonidentity that is reduced and mastered by the identity term of the *positive* dialectic in its traditional form.[6] As Foucault remarks in an early essay, "dialectics does not liberate differences; it guarantees, on the contrary, that they can always be recaptured. The dialectical sovereignty of similarity consists in permitting differences to exist, but always under the sign of the negative, as an instance of non-being."[7]

It is not enough, then, to hold on to the concept of the human and then simply embed it dialectically in networks of symbolic, discursive, and material production, for doing so would simply reenact the retreat-and-return of the subject-as-origin that gave rise in the first place to Foucault's brilliant dismantling of this maneuver in essays like "Nietzsche, Genealogy, History." It means, rather, rethinking the notion of the human *tout court*—a project that fields outside of cultural and social theory have been vigorously engaged in over the past two decades. In recent work in cognitive ethology, field ecology, cognitive science, and animal rights philosophy, for instance, it has become abundantly clear that the humanist habit of making even the *possibility* of subjectivity coterminous with the species barrier is deeply problematic.[8] This body of work has pursued the dismantling of humanism from a direction diametrically opposed to that of Foucault; instead of eroding the boundary between the human subject and its networks of production, it has taken the conceptualization of humanist subjectivity at its word and then shown how humanism must, if rigorously pursued, generate its own deconstruction, once the traditional marks of the human (reason, language, tool use) are found beyond the species barrier. Donna Haraway summarizes many of these developments in her groundbreaking "A Cyborg Manifesto." "By the late twentieth century in United States scientific culture," she writes,

> the boundary between human and animal is thoroughly breached. The last beachheads of uniqueness have been polluted, if not turned into amusement parks—language, tool use, social behavior, mental events. Nothing really convincingly settles the separation of human and animal.... Movements for animal rights are not irrational denials of human uniqueness; they are clear-sighted recognition of connection across the discredited breach of nature and culture.[9]

It should not be assumed, however, that the ethical and political stakes in this boundary erosion are limited to the well-being of nonhuman animals alone. Indeed, the imperative of posthumanist critique may be seen from this vantage—and *is* seen by thinkers like Haraway—as of a piece with larger liberationist political projects that have historically had to battle against the strategic deployment of humanist discourse *against other human beings* for the purposes of oppression. Humanism, in short, is a *discourse;* it is species-specific in its logic (which rigorously separates human from nonhuman) but not in its effects (it has historically been used to oppress both human and nonhuman others). As Gayatri Spivak points out,

> the great doctrines of identity of the ethical universal, in terms of which liberalism thought out its ethical programmes, played history false, because the identity was disengaged in terms of who was and who was not human. That's why all these projects, the justification of slavery, as well as the justi-

fiction of Christianization, seemed to be alright; because, after all, these people had not graduated into humanhood, as it were.[10]

In this light, it is understandable that traditionally marginalized groups and peoples would be loath to surrender the idea of full humanist subjectivity, with all of its privileges, at just that historical moment that they seem poised to "graduate" into it. But, as a host of theorists and critics of contemporary society have pointed out, it is not as if we have a choice about the coming of posthumanism; it is *already* upon us, most unmistakably in the sciences, technology, and medicine. Haraway has argued as forcefully as anyone that our current moment is irredeemably posthumanist because of the boundary breakdowns between animal and human, organism and machine, and the physical and the nonphysical ("Manifesto" 151–55)—a triple hybridity that we can find readily exemplified any evening on cable television, as in a recent program on the U.S. Navy's Marine Mammal project, in which highly trained bottlenose dolphins (human/animal) are fitted with video apparatuses (organism/machine) to locate underwater objects and beam their location back on the Cartesian grid of satellite mapping (physical/nonphysical).

For Haraway, the ethical and political implications of this sort of "cyborg" posthumanism are extremely ambivalent and totally inescapable. "From one perspective," she writes,

> a cyborg world is about the final imposition of a grid of control on the planet.... From another perspective, a cyborg world might be about lived social and bodily realities in which people are not afraid of their joint kinship with animals and machines, not afraid of permanently partial identities and contradictory standpoints. The political struggle is to see from both perspectives at once because each reveals both dominations and possibilities unimaginable from the other vantage point. ("Manifesto" 154)

Not surprisingly, then, the humanist avoidance of the posthumanist imperative to "see from both perspectives" has, as Latour has pointed out, definite *pragmatic* ramifications. Humanist modernity, he argues, is predicated on a kind of paradox. Even as modernity "creates mixtures between entirely new types of beings, hybrids of nature and culture," it "creates two entirely distinct ontological zones: that of human beings on the one hand; that of nonhumans on the other" (10–11). For Latour, this structure has the pragmatic payoff of enabling humanist modernity to "innovate on a large scale in the production of hybrids"—in the production, for example, of genetically engineered organisms like the aggressively marketed OncoMouse for cancer research—because the "absolute dichotomy between the order of Nature and that of Society" prevents the question of the dangerous mixture of ontological categories from ever arising (40, 42). But if the modernist constitutional separation of human and nonhuman has the practical advantage of allowing the proliferation of hybrid networks, it also has the pragmatic drawback (as the strategy of repression always does) of ill equipping contemporary society to explore in a thoughtful way how its relations to and in hybrid networks should be lived.

To do *that*, we must, Latour argues, move beyond the humanist constitution and rethink the notion of politics itself. "The political task starts up again, at a new cost," he writes. "It has been necessary to modify the fabric of our collectives from top to bottom in order to absorb the citizen of the eighteenth century and the worker of the nineteenth. We shall have to transform ourselves just as thoroughly in order to make room, today, for the nonhumans created by science and technology," for "so long as humanism is constructed through contrast with the object that has been abandoned to epistemology, neither the human nor the nonhuman can be understood" (136). But this posthumanist politics that a posthumanist epistemology can help make possible is not, as Haraway reminds us, a matter of voluntarism; it is not as if having a good attitude and taking thought will restore a hybridized world to the clarity and definiteness of identity for the purposes of political praxis. Indeed, as Haraway points out,

> Most important obligations and passions in the world are unchosen; "choice" has always been a desperately inadequate political metaphor for resisting domination and for inhabiting a livable world. Interpellation is not about choice; it is about insertion.... If technological products are cultural actors, and if "we," whoever that problematic invitation to inhabit a common space might include, are technological products at deeper levels than we have yet comprehended, then what kind of cultural action will forbid the evolution of OncoMouse™ into Man™?[11]

Feminist Philosophy of Science and the Detour of "Objectivity"

One of the most prominent and important attempts to answer Haraway's question—and to pursue more generally the prospect of posthumanist theory—has been undertaken by feminist philosophy of science, which has sought to ground "cultural action" by attempting to rehabilitate the notion of objectivity. What is interesting about this desire for "objectivity" is that it issues from a line of critique that has reminded us again and again that putatively "objective" scientific accounts are just as socially constructed as any other, and, moreover, that what we might call the ideology of objectivity has typically operated much to the detriment of women and other marginalized people. In a passage justly famous for its candid statement of the contradictory theoretical desires that characterize much feminist philosophy of science, Haraway writes:

> I think my problem and "our" problem is how to have *simultaneously* an account of the radical historical contingency for all knowledge claims and knowing subjects, a critical practice for recognizing our own "semiotic technologies" for making meanings, *and* a no-nonsense commitment to faithful accounts of a "real" world, one that can be partially shared and friendly to earth-wide projects of finite freedom, adequate material abundance, modest meaning in suffering, and limited happiness.[12]

There are several issues on the table in this passage that are crucial for refiguring the relationship between knowledge, ethics, and political praxis for feminism and beyond. But what is most important for my purposes is the linkage between the ethical and political values cataloged at the end of the passage and the "faithful" accounts of a "real world" that should underwrite or otherwise serve as a foundation for the practice of those values. This strategy in Haraway, Evelyn Fox Keller, Sandra Harding, and others, I now want to argue, is counterproductive because it thrusts the discussion back into a representationalist frame that is both epistemologically inadequate to the task at hand and potentially troubling both politically and ethically.[13]

I agree wholeheartedly with Haraway that "the projects of crafting reliable knowledge about the 'natural' world cannot be given over to the genre of paranoid or cynical science fiction," that "social constructionism cannot be allowed to decay into the radiant emanations of cynicism" ("Situated Knowledges" 184), so that, for example, what counts as knowledge is determined by nothing more than which laboratory has the most money. But I wholeheartedly *disagree* that this means we should redouble our commitment to what Harding has called "strong objectivity"—a leaner if not meaner scientific method that would "identify and eliminate distorting social interests and values from the results of research" by "systematically examining all of the social values shaping a particular research process."[14] The problem with Harding's position, of course, is that it assumes that there *is* some space from which to survey our "social interests and values" without at the same time being bound by those interests and values—a space, in other words, of noncontingent observation, a place where one can tally up all of the "blind spots" without having that tally compromised, rendered less than "objective," by its own blind spot. Even if Harding wants to break with an "absolute" sense of objectivity that presumes what Richard Rorty calls "a God's-eye standpoint," a "view from nowhere," she does so only to rely on a "procedural" form of objectivity that assumes that the chaff of "distorting social interests and values" can be objectively separated from the wheat of nondistorting ones.[15] And when one asks, "distorting in relation to *what*?" then it seems (as the ocular figuration of the problem suggests) that we are back within the representationalist frame, which fails to acknowledge what the *other* half of Harding, the constructivist half, knows full well: that there is "no way," as Rorty puts it "of formulating an independent test of accuracy of representation—of reference or correspondence to an 'antecedently determinate' reality—no test distinct from the success which is supposedly explained by this accuracy" (*ORT* 6). To use the terms of Francisco Varela, Evan Thompson, and Eleanor Rosch in *The Embodied Mind*, Harding's "strong objectivity" is in the end just a form of "weak representationalism"—representationalism with apologies, as it were—because, in saying "that different perceiving organisms simply have different perspectives on the world," it "continues to treat the world as pregiven; it simply allows that this pregiven world can be viewed from a variety of vantage points."[16]

   Let me say again that my intention is not to take issue with the admirable *political* values and aims of Harding's project: to argue against the uses of science in promoting inequality and environmental degradation; to critique the reproduction of Eurocentrism and racism in scientific institutions; to open the practice and resources of science as an institution and a discipline to marginalized peoples. My point, rather, is to underline the theoretical incoherence of presuming that these values and aims must be grounded in some notion of objectivity. Just as Haraway paradoxically insists that "only partial perspective promises objective vision" ("Situated Knowledges" 190), so Harding argues that "the systematic activation of democracy-increasing interests and values—especially in representing diverse interests in the sciences when socially contentious issues are the object of concern—in general contributes to the objectivity of science" (18). In response to which one must simply say, it does not follow. How can greater diversity of socially "interested" knowledges add up to a more "objective" sort of knowledge, when objectivity is by definition precisely the sort of knowledge you get once you have removed, rather than expanded, the influence of "social interests and values" on it? Harding might respond that only "distorting" and "obscuring" interests and values need be removed, but that, indeed, is precisely the rub, for who is to say—*especially* without foreclosing the sort of democratic debate and radical questioning that Harding rightly encourages—when an interest is distorting and when it is not? And if I make that claim of distortion about another, then how is *my* interest not unduly influencing the process?

These difficulties are symptomatic of the essential fallacy at work in the assumption, to borrow Barbara Herrnstein Smith's characterization,

> that objectivism is wrong when practiced by the wrong people for the wrong reasons, but right when practiced by the right people for the right reasons: specifically, that objectivist arguments are culpably "authoritarian" when they issue from powerful agents attempting to justify their own self-interested actions, but laudably "critical" when they issue from disinterested agents exposing the unjust acts of powerful people against subordinated people. Such distinctions, however, are impossible to maintain either theoretically or practically.[17]

Although we have already lingered on the theoretical incoherence diagnosed by Smith, what is not so clear—but what is every bit as important—are the disabling practical implications mentioned by Smith; for the assumption that there is a necessary correlation between the legitimacy or achievement of the political aims of feminist philosophy of science and the attainment of objectivity ("strong" or not) on the epistemological plane is, I think, rhetorically counterproductive, because it creates a self-defeating contradiction between Harding's polemical, political project (to open up scientific knowledge to "outsider" values and perspectives) and her theoretical, epistemological project (to continue to define what counts as legitimate knowledge by measuring it against a representationalist standard of "objectivity"). To put it another way, Harding's polemical/political project wants to open up science as an institution to social representation, but her theoretical and epistemological premium on "objectivity"—in separating social interests and values from the objects of research, in separating distorting from nondistorting values—only reinforces the disciplinary *insularity* of science as a discursive community from the rest of social discourse.

"Democratic values," Harding writes, "ones that prioritize seeking out criticisms of dominant belief from the perspective of the lives of the least advantaged groups, tend to increase the objectivity of the results of research" (18). But how can such a thing logically be the case? What *is* the case, however, is that such a process, although it has nothing to do with objectivity—except maybe laudably calling that very notion into ever more radical question—*does* have plenty to do with politics. "Representing diverse interests" in the sciences and "seeking out criticisms of dominant belief" in the sciences *do just that;* they do not "achieve the elimination of objectivity-damaging social values and interests" but instead *propagate* those values and interests for the purposes of greater democratic representation of the points of view in the knowledge-making process of Harding and those she presumes to speak for. And that, from a pragmatic point of view, is all that a social and political critique of knowledge *can* do. In this light (it probably goes without saying), the practical, rhetorical disadvantages of the "strong objectivity" program I have just noted take on renewed significance.

As a range of self-professedly "relativist" theorists have pointed out, it need not be assumed that alternative political and ethical visions must be "grounded" in some objective view of the world, and that only by reference to such objectivity does one have the right to criticize the existing order of things. After all (to tamper a bit with Haraway's formulation), what if it turns out that our objective "faithful accounts of a 'real' world" turn out "objectively" *not* to be ones "that can be partially shared and friendly to earth-wide projects"? What do we do then, abandon those projects? Certainly not. In fact, as Malcolm Ashmore, Derek Edwards, and Jonathan Potter have argued, "Realism is no more secure than relativism in making sure the good guys win, nor even in defining who the good guys are—except according to some specific realist assumptions that place such issues outside of argument." For them, it is the objectivist position that courts political conservatism and quietude, while "it is for relativists and constructionists that the good life is to be lived and *made*, as and in accountable social action, including that of social analysis."[18] Indeed, from Barbara Herrnstein Smith's point of view, there may be "a certain grandeur" to objectivist claims, but "What is sacrificed to obtain that grandeur, however—namely, the acknowledgment of both human variability and the mutability of the conditions of human existence—is likely to be paid sooner or later in political coin" (292).

To attempt to ground progressive political praxis in objectivity is thus—to borrow Rorty's phrase about Habermas—to "scratch where it does not itch" by attempting to provide a metanarrative of objectivity, rationality, or universalism to ground the contingent "first-order" narratives at work in social life.[19] As we have seen, "The pragmatist's justification of toleration, free inquiry, and the quest for undistorted communication can only take the form of a comparison . . . not by reference to a criterion"—such as objectivity—"but by reference to various detailed practical advantages" (*ORT* 29). This does not mean, as the archrealist or even representationalist-with-apologies is sure to rush in and exclaim, that the pragmatist or relativist has no way to show us that the "real" world exists; for this charge, as Ashmore et al. point out, "trades upon the objectivist assumption that rejecting realism is the same thing as rejecting everything that realists think is real" (8). As we have already seen, the pragmatist, as Rorty explains, "believes, as strongly as does any realist, that there are objects which are *causally* independent of human beliefs and desires" (*ORT* 101), but in granting this "causal stubbornness" to the world she does not grant the "real" or the "object" "an *intentional* stubbornness, an insistence on being *described in a certain way*, its *own* way" (*ORT* 83).

This does not mean that so-called facts of the sort invoked by feminist philosophy of science's realist side are then simply ad hoc constructions driven only by political expediency. It is true that from a pragmatist point of view nothing prevents us *epistemologically* from going around and making up knowledge claims that seem upon reflection outlandish; but much prevents us *institutionally* and pragmatically from doing so if we want those claims to receive a serious hearing and count as knowledge within a given discourse.[20] As Rorty explains:

> Facts are hybrid entities; that is, the causes of the assertability of sentences include both physical stimuli and our antecedent choice of response to such stimuli. To say that we must have respect for the facts is just to say that we must, if we are to play a certain language game, play by the rules. To say that we must have respect for unmediated causal forces is pointless. It is like saying that the blank must have respect for the impressed die. The blank has no choice, and neither do we. (*ORT* 81)

The pragmatic critique, then, does not say that the "real world" does not exist, that there is no such thing as a "fact," or that we can blithely falsify the data as we go along. It simply means that we should jettison the epistemological pretensions that want to ground certain practices and values in "objectivity" and ground them instead in whether or not they work, as agents of adaptation to an environment, for contingent, revisable purposes. Thus, "from a pragmatist point of view," Rorty writes,

> to say that what is rational for us now to believe may not be *true*, is simply to say that somebody may come up with a better idea. It is to say that there is always room for improved belief, since new evidence, or new hypotheses, or a whole new vocabulary, may come along. For pragmatists the desire for objectivity is not the desire to escape the limitations of one's community, but simply the desire for as much intersubjective agreement as possible. (*ORT* 23)

On this view, it is perfectly possible to appeal to experimental evidence (as many antirealists do all the time) not because it provides a more "accurate" or transparent reflection of the way things "really are" in the world, but rather because it is persuasive within the rules of knowledge for a given discourse. (And those rules, of course, will vary in breadth, stringency, and prescriptiveness from discipline to discipline.)

The objectivist/realist will no doubt want to challenge this claim by appealing to science's effectivity, but it is quite possible to account for that effectivity by extending the powerful social constructivist arguments mobilized by Latour, Steve Woolgar, and others. Science, on this view, is privileged not because of its representational transparency to the real, but rather because it *works*. And this fact, in turn—despite the realist attempt to use science's effectivity as evidence of the freedom of science's truth claims from the arena of social power and political rhetoric—only foregrounds the imbrication of science in that very arena, for the question we then must ask is, "works for *whom*, for what purposes?" In this context, it makes sense, of course, that feminist philosophy of science would want to trade upon the considerable rhetorical power of "objectivity" to affect social and institutional change. But the problem, as we have seen, is that these claims for "objectivity" are made not within a rhetorical, political frame—in which one cunningly appropriates notions that are philosophically suspect because they carry powerful appeal for specific audiences (other feminists, say, who are not philosophers or literary theorists)—but are offered instead squarely within an epistemological register, as a theory about the status of knowledge claims. And if one then wants to ask "so who cares about epistemology?" (which is, after all, a pragmatist question), the answer must be that *we* care—and so do Haraway, Harding, and Fox Keller, who, after all, write epistemological books for theoretical, academic audiences. If we want to meet the epistemological critique of objectivity by devaluing epistemology itself as being "academic" in the worst sense, we must remember Ashmore et al.'s reminder: "But *we* are academics, for whom it is proper, even essential, to care about the epistemic and ontological status of claims to knowledge" (9).

What I am suggesting, then, is that the pragmatic and political dimension of recent feminist philosophy of science can and should be disengaged from the objectivist epistemological pretensions that undercut its political and ethical commitment. In that disengagement, it can then theorize more coherently the desire that it voices again and again: for a contingency that is not myopic, a constructivism that is more than just self-serving stories. It could be argued by Haraway, Harding, et al. in response to this suggestion that the second-order cybernetics we are about to examine cannot very well critique the use of "objectivity" and at the same time offer *itself* as a transdisciplinary paradigm that claims universal descriptive validity. But the rejoinder, as we shall see, is that if we agree that *all* critiques or theories are reductive of difference (because they are all contingent, which means that we could have described things otherwise), then the issue becomes how to build a confrontation with that fact into the epistemology one is using, rather than continuing to pretend that this contingency does not exist by strategically repressing it.

In the meantime, to avoid constantly undercutting their political critique with an epistemology ill equipped to serve it, when Haraway in "Situated Knowledges" says "objectivity" she should instead say what she really means, which is "situatedness" and "responsibility," and when Harding says "objectivity" she should instead just say "democracy" and "representation of marginalized voices." This will be difficult for feminist philosophy of science to do, because it is, after all, philosophy of *science*. But once it has affected this disengagement, it will have much to teach pragmatists like Rorty, whose complacent ethnocentrism, as we have already seen, needs to be confronted with the more muscular pragmatism that is alive and well in Haraway, Harding, and Fox Keller, the latter of whom—in her study *Secrets of Life, Secrets of Death*—puts squarely on the front burner the sort of question often avoided or blithely glossed over by pragmatism in its Rortyan incarnation. "From critical theory, to hermeneutics, to pragmatism," she writes,

> the standard response to so-called relativist arguments has been that the scientific stories are different from other stories for the simple reason that they "work." If there is a single overriding point I want to make...it is to identify a chronic ellipsis in these responses: As routinely as the effectiveness of science is invoked, equally routine is the failure to go on to say what it is that science works *at*, to note that "working" is a necessary but not sufficient constraint.[21]

Only by forcing examination of these *specific, material* effects of scientific discourse and practice can we forge a more socially and politically responsive pragmatist critique of knowledge that understands that if science is "what works," it always works *at* something *for* a "particular 'we'...embedded in particular cultural, economic, and political frames" (Fox Keller, *Secrets of Life* 5). Only by paying this sort of attention can we force the pragmatist commitment to contingency to be true to its word and undertake a full critique of what Fox Keller calls the "romance of disembodiment" on not only epistemological grounds, but political ones as well.

When Loops Turn Strange: From First- to Second-Order Cybernetics
In light of the posthumanist imperative I have been invoking thus far, systems theory has much to offer, I believe, as a general theoretical orientation. Unlike feminist

philosophy of science, it does not cling to debilitating representationalist notions. And unlike Enlightenment humanism in general, its formal descriptions of complex, recursive systems are not grounded in the figure of "Man" and in the dichotomy of human and nonhuman. Indeed, in light of the posthumanist context I have sketched here, the signal virtue of systems theory is, as Dietrich Schwanitz puts it, that it has "progressively undermined the royal prerogative of the human subject to assume the exclusive and privileged title of self-referentiality (in the sense of recursive knowledge about knowledge)."[22]

The promise and power of systems theory reside not only in its posthumanism, however, but also in its ability to offer a much more rigorous and coherent way to theorize the extraordinarily complex "hybrid" or "cyborg" networks of the sort described in much of Haraway's work, and by Latour in the opening pages of *We Have Never Been Modern*. Recounting the experience of reading a newspaper article on the ozone layer, Latour observes:

> The same article mixes together chemical reactions and political reactions. A single thread links the most esoteric sciences and the most sordid politics, the most distant sky and some factory in the Lyon suburbs, dangers on a global scale and the impending local elections or the next board meeting. The horizons, the stakes, the time frames, the actors—none of these is commensurable, yet there they are, caught up in the same story. (1)

What suggests a privileged place for systems theory, then, in meeting the theoretical challenges posed by the cyborg hybridity of postmodern society is its ability to mobilize the same theoretical apparatus across domains and phenomena traditionally thought to be pragmatically discrete and ontologically dissimilar, *while at the same time* offering (as we shall see with recent work on "the observation of observation") a coherent and compelling account of the ultimate contingency of any interpretation or description.

Moreover, systems theory has retained, from its very genesis, a crucial *pragmatic* dimension that distinguishes it from other theoretical approaches (deconstruction, for example) with which it otherwise has much in common. As Latour points out, the reason it is not adequate to charge the sort of approach taken by systems theory with talking about only "meaning effects and language games," is that here, "these are really at stake, but in a new form that has a simultaneous impact on the nature of things and on the social context, while it is not reducible to one or the other" (5). For example (as one of the central figures of early cybernetics, Norbert Wiener, reminds us), much of the pioneering work in the field was centered on the problem of improving targeting mechanisms for antiaircraft artillery in World War II.[23] In such instances, we are not dealing only with language games, but with (to use Latour's plangent phrase) "arrangements that can kill us all" (5).

I will return to these issues in more detail, but for now we need to keep in mind that there exists within systems theory itself an important distinction between first- and second-order cybernetics that we will need to understand before we can grasp the full originality and importance of the work of Humberto Maturana, Francisco Varela, and Niklas Luhmann. To get a sense of *first*-order cybernetics and why posthumanist theory must move beyond it, there is no more instructive example than the cultural anthropologist and intellectual polymath Gregory Bateson, who from the 1940s through the 1970s engaged in an ambitious and wide-ranging attempt to extend the new theoretical model of systems theory and cybernetics, first developed in information engineering and biology, into the social sciences to describe the basic formal dynamics of alcoholism, communication among wolves and dolphins, primitive art and ritual, ecological crisis, schizophrenia, and much else besides.

Bateson's work embodies many of systems theory's basic theoretical assumptions and commitments; above all, it is *integrative* or *organizational*, as opposed to analytical and atomistic. As one of the founding figures of systems theory, Ludwig von Bertalanffy, writes:

> While in the past, science tried to explain observable phenomena by reducing them to an interplay of elementary units investigatable independently of each other, conceptions appear in contemporary science that are concerned with what is somewhat vaguely termed "wholeness," i.e., problems of organization, phenomena not resolvable into local events, dynamic interactions manifest in the difference of behavior of parts when isolated or in a higher configuration, etc.; in short, "systems" of various orders not understandable by investigation of their respective parts in isolation.... There appear to exist general system laws which apply to any system of a certain type, irrespective of the particular properties of the system and of the elements involved.[24]

As Robert Lilienfeld points out—and as Bertalanffy's formulation suggests—this emphasis on systemic integration and wholeness in systems theory can take one of two forms (and sometimes both alternately in the same thinker, as is the case with Bateson): either the "contextualism" (what we might call the "conventionalism") of the sort that I have been emphasizing throughout this study, or an "organicism" that is less resolutely constructivist and more insistently realist. "The contextualist," Lilienfeld writes, "uses the category of integrating structures (contexts) to explain experience, but denies to these integrating structures any reality of significance. The organicist maintains that [they] . . . are more numerous, coherent, and 'real' than the contextualist wants to admit."[25] For the contextualism that, as Lilienfeld reminds us, links systems theory rather directly to the pragmatism of Peirce and James, "The world is seen as an unlimited complex of change and novelty, order and disorder," which is organized by certain contexts, by "organizing gestalts or patterns," that give meaning to what would otherwise be an unpatterned "noise" of detail (9). As Bateson characterizes it in a particularly instructive discussion:

> the word "idea," in its most elementary sense, is synonymous with "difference." Kant, in the *Critique of Judgment*—if I understand him correctly—asserts that the most elementary aesthetic act is the selection of a fact. He argues that in a piece of chalk there are an infinite number of potential facts. The *Ding an sich*, the piece of chalk, can never enter into communication or mental process because of this infinitude. The sensory receptors cannot accept it; they filter it out. What they do is to select certain *facts* out of the piece of chalk, which then become, in modern terminology, information.
>
> I suggest that Kant's statement can be modified to say that there is an infinite number of *differences* around and within the piece of chalk. There are differences between the chalk and the rest of the universe, between the chalk and the sun or the moon. And within the piece of chalk, there is for every molecule an infinite number of differences between its location and the locations in which it *might* have been. Of this infinitude, we select a very limited number, which become information.[26]

As Bateson is fond of saying (invoking Korzybski's famous dictum), *"the map is not the territory"* (*Steps* 449; emphasis in the original); the sort of knowledge (or information) you get depends on the context (or code) you deploy, and not—here we should remember Rorty's critique of representationalism—on a more or less transparent reflection of the "substance" of the object being described. As Lilienfeld points out, what this means—and again this situates systems theory within a broader philosophical pragmatism—is that "From the assumptions of contextualism a specific theory of truth emerges—operationalism. . . . Truth is 'the successful working of a idea' within a specific (and always limited) context. Truth is verification in practice" (10).

Moving from the general epistemological orientation of systems theory to the more specific features of its explanatory model, we do well to consult Bateson's work once more. "When we talk about the processes of civilization, or evaluate human behavior, human organization, or any biological system," Bateson writes, "we are concerned with self-corrective systems. Basically these systems are always *conservative* of something. As in the engine with a governor, the fuel supply is changed to conserve—to keep constant—the speed of the flywheel, so always in such systems changes occur to conserve the truth of some descriptive statement, some component of the *status quo*" (*Steps* 429). In Bateson's view, the "essential minimal characteristics of a system"—be it biological, mechanical, or social—are (1) that the system operates upon *differences*, deviations from a norm or baseline that are processed as information; (2) that it consists of "closed loops or networks of pathways along which differences and transforms of differences shall be transmitted" (as when a thermostat detects the difference between its setting and the room temperature, activating the furnace to restore the total loop of room/furnace/thermostat to the desired homeostatic state); (3) that "many events in the system shall be energized by the respondent part rather than by impact from the triggering part" (a principle most clear, perhaps, in phenomena such as color vision, and in the various tricks and demonstrations, such as the parallax effect, which show how the nervous system actively and constructively responds to environmental stimuli rather than simply registering them in a linear fashion); (4) that systems "show self-correctiveness in the direction of homeostasis and/or in the direction of runaway" (*Steps* 482).

This last characteristic always involves the fundamental principle of cybernetics: circular causality or "recursivity," a principle whose most well known example probably remains the "feedback loop," of which there are two types: negative feedback, in which information is processed by the system in such a way as to maintain the harmony, homeostasis, or directionality of the system, and positive feedback, in which information is processed in a such a way as to destabilize the system and create what is sometimes called a "vicious cycle" (what Bateson calls "runaway").

Although positive feedback is important to recent work in complexity theory,[27] we must leave it aside to concentrate on negative feedback, a famous example of which is offered by Steve J. Heims in his social history of cybernetics: "A person reaches for a glass of water to pick it up, and as she extends her arm and hand is continuously informed (negative feedback)—by visual or proprioceptive sensations—how close the hand is to the glass and then guides the action accordingly, so as to achieve the goal of smoothly grabbing the glass" (15). What is immediately intriguing about this example of negative feedback—and about the principle of circular causality in general—is that it contains a paradox, one that second-order cybernetics will pursue to its logical conclusions, as first-order cybernetics never really did: A causes B *and* B causes A. As Heims explains, "The process is circular because the position of the arm and hand achieved at one moment is part of the input information for the action of the next moment" (15–16). And hence, the system is characterized by "recursivity," which, as Niklas Luhmann defines it, is a process that "uses the results of its own operations as the basis for further operations—that is, what is undertaken is determined in part by what has occurred in earlier operations. In the language of systems theory...one often says that such a process uses its own outputs as inputs."[28]

Despite its interdisciplinary range and explanatory power, Bateson's work stops short of pursuing the full implications of this paradoxical fact about recursivity (A causes B *and* B causes A), and the contingency of all observation to which such paradoxicality attests (we can say *either* A causes B *or* B causes A; thus it is always possible to observe otherwise). The move from first- to second-order cybernetics is characterized, as Heinz von Foerster argues in *Observing Systems*, by the full disclosure of this fundamental epistemological problem:

> (i) Observations are not absolute but relative to the observer's point of view (i.e. his coordinate system: Einstein); (ii) Observations affect the observed so as to obliterate the observer's hope for prediction (i.e. his uncertainty is absolute: Heisenberg).
>
> After this, we are now in the possession of the truism that a description (of the universe) implies one who describes (observes it).[29]

What is most intriguing about Bateson's work is that, on the one hand, he wants to insist, in essays like "Redundancy and Coding" and "Cybernetic Explanation," on the contingency of observation, on the constructivist point that the sort of knowledge you get depends on the code or map that you use—that "the map is not the territory." On the other hand, that recognition of contingency gets undone by Bateson's totalizing insistence that there is a single, total loop or overarching "*pattern which connects*" observer and observed,[30] so that what looks at first glance like contingent observation is instead determined "from behind" by the total pattern of existence, generating what Bateson calls an immanent "mental determinism" (*Steps* 465). To invoke the categories we borrowed earlier from Lilienfeld, we may say that here, Bateson's contextualism is undercut by his organicism. This is quite clear in later essays like "Form, Substance, and Difference," where Bateson writes:

> The cybernetic epistemology which I have offered you would suggest a new approach. The individual mind is immanent but not only in the body. It is immanent also in pathways and messages outside the body; and there is a larger Mind of which the individual mind is only a subsystem. This larger Mind is comparable to God and is perhaps what some people mean by God,

but it is still immanent in the total interconnected social system and plane-
tary ecology. (*Steps* 460)

It is at moments like these that the epistemological rigor of second-order cybernetics proves decisive and invaluable. As von Foerster suggests, the
crucial realization of second-order cybernetics is that you cannot do justice to constructivist contingency—regardless of whatever liberating ethical or political implications that might flow from it—and at the same time hold (in "organicist" fashion) that the efficacy of your description is that it is more or less transparent to the
total "pattern" or "order" of existence; for once it is acknowledged that observation
is contingent (i.e., could be otherwise), then it must also be acknowledged that total
loops such as those imagined by Bateson must always turn into "strange" loops of
the sort imaged by M. C. Escher's Möbius strip. As Ranulph Glanville and Francisco Varela remind us in their elegant little demolition of total loops titled "Your
Inside Is Out and Your Outside Is In (Beatles, [1968])," the distinction between inside and outside, system and environment, mind and nature, always contains a paradox that makes the distinction turn back upon itself to form a "strange" loop. This
is so, they argue, because when we draw any putatively final distinction in either intension or extension—when we attempt to distinguish either the elementary or the
universal—"we require that its distinction has no inside and, at the same time we
place, in this non-existent inside, a further distinction which asserts that the distinction of the fundamental was the last distinction!"[31] Thus, they continue, "at the
extremes we find there are no extremes. The edges dissolve BECAUSE the forms are
themselves continuous—they re-enter and loop around themselves" (640), not like
a Batesonian circle of the total system but like a Möbius strip, a more fitting image
for the paradoxicality of distinction—a paradoxicality that, second-order cybernetics forces us to say, *must always accompany the assertion of the contingency of the observer,
of the fact that an observation could always be otherwise.*[32]

This abandonment of the total "pattern which connects" on be-
half of the contingency of observation (and the sort of systemic heterogeneity it
makes recognizable) links second-order cybernetics rather directly to broader currents of postmodern theory of the sort practiced by Gilles Deleuze and Félix Guattari. "I part company with Bateson," Guattari writes,

> at the point where he defines action and enunciation as mere segments of
> the ecological sub-system known as context.... There is no overall hierar-
> chy of enunciative ensembles and their sub-sets, whose components can be
> located and localized at particular levels. Those ensembles are made up of

heterogeneous elements which acquire consistency and persistence only as they cross the thresholds that bound and define one world against another. They are . . . [like] Schlegel's "little works of art" ("Like a little work of art, a fragment has to be totally detached from the surrounding world and closed upon itself like a hedgehog")[33]

—or, as we are about to see, like the autopoietic organizations of second-order cybernetics, which—far from participating in an "immanent determinism" driven by the total "pattern which connects"—are totally self-referential because they exist by virtue of what Maturana and Varela will call their "operational closure." Under the sign of second-order cybernetics and its postmodern cognates, knowledge appears instead, in Varela's words,

> more and more as built from small domains, that is, microworlds and microidentities. . . . [S]uch microworlds are not coherent or integrated into some enormous totality regulating the veracity of the smaller parts. It is more like an unruly conversational interaction: the very presence of this unruliness allows a cognitive moment to come into being according to the system's constitution and history.[34]

Here, then, we glimpse the full implications of second-order cybernetics' emphasis on the contingency of observation, its constant reminder, as Maturana and Varela put it, that "everything that is said is said by someone." Because all contingent observations are made by means of the "strange loop" of paradoxical distinction between inside and outside, x and not-x, "every world brought forth necessarily hides its origins. By existing, we generate cognitive 'blind spots' that can be cleared only through generating new blind spots in another domain. We do not see what we do not see, and what we do not see does not exist."[35]

Between the Scylla of Realism and the Charybdis of Idealism:

Autopoiesis and Beyond

The key distinction for the theory of autopoiesis (or "self-production") as articulated by Maturana and Varela—the distinction that (as we shall see in a moment) allows its decisive conceptual innovation, its account of systems that are both open and closed—is the distinction between "organization" and "structure." As they explain it, "*Organization* denotes those relations that must exist among the components of a system for it to be a member of a specific class"; it is that which "signifies those relations that must be present in order for something to exist." *Structure*, on the other hand, "denotes the components and relations that actually constitute a particular unity and make its organization real" (*Tree* 46, 47). For example, the basic and necessary *organization* of the water-level regulation system in a toilet consists of a float and a bypass valve. But in terms of the *structure*, the float that is made of plastic could be replaced by one made of wood "without changing the fact," as Maturana and Varela somewhat infelicitously put it, that there would still be "a toilet organization" (*Tree* 46). This basic distinction between organization and structure will mark a crucial epistemological innovation in their attempt, as they put it, to "walk on the razor's edge, eschewing the extremes of representationalism (objectivism) and solipsism (idealism)" (*Tree* 241). It will also, more broadly, enable a reconceptualization of the relationship between *system* (organization + structure) and *environment* (everything outside the system's boundaries) that will mark a definitive break with the first-order cybernetics of Bateson.

For Maturana and Varela, what characterizes all living things is that they are "*autopoietic organization[s]*," that is, "they are continually self-producing" (*Tree* 43) according to their own internal rules and requirements. In more general terms, what this means is that all autopoietic entities are *closed* — or, to employ Niklas Luhmann's preferred term, "self-referential" — on the level of *organization*, but *open* to environmental perturbations on the level of *structure*. This is clearest, perhaps, in Maturana and Varela's contention that all autopoietic entities are defined by "*operational closure.*" "It is interesting to note," they write,

> that the operational closure of the nervous system tells us that it does not operate according to either of the two extremes: it is neither representational nor solipsistic.
>
> It is not solipsistic, because as part of the nervous system's organism, it participates in the interactions of the nervous system with its environment. These interactions continuously trigger in it the structural changes that modulate its dynamics of states. . . .
>
> Nor is it representational, for in each interaction it is the nervous system's structural state that specifies what perturbations are possible and what changes trigger them. (*Tree* 169)[36]

Environmental "triggers" and "perturbations," then, take place on the level of structure, but what may be *recognized* as a perturbation or trigger is specified by the entity's organization and operational closure. What this means, Maturana and Varela conclude squarely against the first-order cybernetics of Bateson, is that the model of the nervous system "picking up information" from the environment is misleading (*Tree* 169); "information," as Varela, Thompson, and Rosch put it in *The Embodied Mind*, is not "a prespecified quantity, one that exists independently in the world and can act as the input to a cognitive system." After all, they ask, "how are we to specify inputs and outputs for highly cooperative, self-organizing systems such as brains?" (139). The difference between cognitive systems (and, Maturana and Varela would argue, autopoietic systems in general) and input-output devices is, in the words of Marvin Minksy, "that brains use processes that change themselves—and this means we cannot separate such processes from the products they produce" (quoted in *Embodied Mind* 139).

Here, then, we can see how second-order cybernetics radicalizes the concept of recursivity abandoned prematurely by first-order cybernetics. As we have seen, first-order cybernetics avoids the crude representationalism and realism that holds, as Richard Rorty puts it, that " 'making true' and 'representing' are reciprocal relations: the nonlinguistic item which makes *S* true is the one represented by *S*" (*ORT* 4). But it does so only to smuggle representationalism back in in the form of the input-output model and the notion of "information processing." For Maturana and Varela, revealing the poverty of the representational frame for making sense of such phenomena as perception, color vision, cognition, and memory is absolutely crucial to their entire epistemological project, which aims to "negotiate a middle path between the Scylla of cognition as the recovery of a pregiven outer world (realism) and the Charybdis of cognition as the projection of a pregiven inner world (idealism)." "These two extremes," Varela et al. contend, "both take representation as their central notion: in the first case representation is used to recover what is outer; in the second case it is used to project what is inner" (*Embodied Mind* 172).

And at this juncture, Maturana and Varela in *The Tree of Knowledge* broach the question that any antirepresentationalist epistemology sooner or later must confront: namely, the question of relativism. "If we deny the objectivity of a knowable world," they ask, "are we not in the chaos of total arbitrariness because everything is possible?" The way "to cut this apparent Gordian knot," they respond, is to realize that the first principle of any sort of knowledge whatsoever is that "everything said is said by someone"—to foreground, in short, the problem of observation (*Tree* 135). As Varela et al. put it, "Our intention is to bypass entirely this logical geography of inner versus outer by studying cognition not as recovery or projection but as embodied action" (172)—"embodied" because cognition depends on the "individual sensorimotor capacities" of the embodier in situ, and "active" (or "enactive") because the cognitive structures that guide perception and action—as dramatically demonstrated by the example of color vision—"emerge from the recurrent sensorimotor patterns that enable action to be perceptually guided" (173).[37] The full definition of "embodiment," then, is a self-referential, self-organizing, and nonrepresentational system whose modes of emergence are made possible by the history of structural coupling between the autopoietic entity and an environment to which it remains closed on the level of organization but open on the level of structure. This cluster of terms constitutes what Varela calls a "radical paradigmatic or epistemic shift" that holds that the lived, concrete, contingent, embodied quality of all knowledge "is not 'noise' that occludes the brighter pattern to be captured in its true essence, an abstraction, nor is it a step toward something else: it is how we arrive and where we stay" ("Reenchantment" 320).

But this acknowledgment of the full complexity of autopoietic systems does not dispense with systematic description altogether. Instead, it recasts the relationship between a system and its elements (or, to use the language of Maturana and Varela, an organization and its structure) as open-ended and yet not random, fundamental and yet not foundational in the usual ontological sense. As Dietrich Schwanitz puts it, "the elements function as units only within the system that constitutes them, they are neither just analytical constructs nor do they rest in some ontological substance. They really do exist, but their existence is only brought about by self-reference and cannot in any way be explained by reference to preexisting ideas, substances or individuals" (272). This loss of meaning (if one wants to put it in that representational way) is, according to Varela, totally unavoidable, and nowhere is this clearer than in his work on perception and cognition, which reveals the temporal structure of the cognitive transition from one moment or action to the next to be extremely "fine" in texture, consisting of a "fast dynamics" or "fast resonance" of neuronal activity in which we find extremely rapid cooperation and competition between distinct neural agents ready to constitute different frames of action and interpretation of the perceptual event. "On the basis of this fast dynamics," Varela explains,

> as in an evolutionary process, one neuronal ensemble (one cognitive subnetwork) finally becomes more prevalent and *becomes the behavioral mode for the next cognitive moment.* By "becomes more prevalent" I do not mean to say that this is a process of optimization: it resembles more a bifurcation or symmetry-breaking form of chaotic dynamics. *It follows that such a cradle of autonomous action is forever lost to lived experience* since, by definition, we can only inhabit a microidentity when it is present, not when it is in gestation. ("Reenchantment" 334; second emphasis mine)

The particular suppleness of this sort of descriptive apparatus, then, is that it provides us with "a philosophical system, a reductive system," as Varela et al. put it, "in which reductive basic elements are postulated as ultimate realities but in which those ultimate realities are not given ontological status in the usual sense" (*Embodied Mind* 118).

More than a few readers have suggested that this way of negotiating the realism/idealism problem constitutes a kind of double-dealing—a cooking of the books of nature, you might say. The Marxist sociologist Danilo Zolo, for example, has suggested that a persistent confusion about the claims and status of autopoiesis haunts the work of Maturana and Varela. On the one hand, Zolo argues, they want to maintain a last, fretful tie to empiricism. They go out of their way to claim that the theory of autopoiesis does not rely on reference to forces or dynamics "not found in the physical universe" (as they put it in *Autopoiesis and Cognition*), that autopoietic unity "is not an abstract notion of purely conceptual validity for descriptive purpose, but is an operative notion."[38] But, on the other hand, they want to espouse a thoroughgoingly constructivist position that holds that any scientific explanation is always, as they put it, "a reformulation of a phenomenon"—that when we describe an autopoietic system, "we project this system upon the space of our manipulations and make a description of this projection" (quoted in Zolo, "Autopoiesis" 67). As Zolo sees it, Maturana and Varela want to hold that predictions about what happens in physical space (as opposed to the abstract and conceptual domain) are valid because, as they put it, "a description, as an actual behavior, exists in a matrix of interactions which (by constitution) has a logical matrix necessarily isomorphic with the substratum matrix within which it takes place" (quoted in ibid., 69). But this, Zolo argues, only redoubles the contradictory status of the claims of autopoiesis. "They forget," Zolo writes,

> that they have already argued that it is impossible to distinguish "between perception and hallucination in the operation of the nervous system";... that nothing can be said about the "substratum" of observation; that knowledge has no object and that everything that can be said is always said by an observer. Thus, it is meaningless to postulate the existence of a "logical isomorphism" between the substratum of the observation and the language of description. (69)

The problem foregrounded but not fully understood, I think, by Zolo's critique—nor, it should be added, is it always clearly articulated by Maturana and Varela—is one we have already mentioned: the problem of *observation*. Maturana offers what is in effect a response to Zolo's critique, and in particular to Zolo's rather fast and loose mobilization of the dichotomies objective/subjective, realist/idealist, and so on:

The fact that science as a cognitive domain is constituted and validated in the operational coherences of the praxis of living of the standard observers as they operate in their experiential domains without reference to an independent reality, *does not make scientific statements subjective*. The dichotomy of objective-subjective pertains to a cognitive domain in which the objective is an explanatory proposition that asserts, directly or indirectly, the operational possibility of pointing to an independent reality. Science does not, and cannot, do that.[39]

But this response only foregrounds the necessity to theorize even more rigorously the concept of observation. "As observers," Maturana and Varela write, "we can see a unity in *different* domains, depending on the distinctions we make"; we can consider the internal states and structures of a system, or we can consider how that system interacts with its environment. For the former observation, "the environment does not exist"; for the latter, "the internal dynamics of that [system's] unity are irrelevant" (*Tree* 135). The key point, then, is that

> both are necessary to complete our understanding of a unity. It is the observer who correlates them from his outside perspective. It is he who recognizes that the environment can trigger structural changes in it. It is he who recognizes that the environment does not specify or direct the structural changes of a system. The problem begins when we unknowingly go from one realm to the other and demand that the correspondences we establish between them (because we see these two realms simultaneously) be in fact a part of the operation of the unity. (*Tree* 135–36)

In his essay "Science and Daily Life," Maturana offers an even more nuanced explanation of his concept of observation, one that helps us to see how Zolo's critique is mounted upon a foundation of epistemological reductionism. In Maturana's view, by contrast, the

> nonreductionist relation between the phenomenon to be explained and the mechanism that generates it is operationally the case because the actual result of a process, and the operations in the process that give rise to it in a generative relation, intrinsically take place in independent and nonintersecting phenomenal domains. This situation is the reverse of reductionism; scientific explanations as generative propositions constitute or bring forth a generative relation between otherwise independent and nonintersecting phenomenal domains, which they thus *de facto* validate. ("Science and Daily Life" 34)

What this means, I take it, is that the scientific explanation or observation constitutes the *relation* between "the phenomenon to be explained" (the observer's view of the system in its environment, which is not possible from the vantage of the system) and the "mechanism" or "operations" (the relation between the system's operationally closed organization and its structure, which is open to environmental triggers).

The key words here, then, are "actual" and "nonintersecting"; the "result of a process" is "actual" not only because it is what the observer *sees*, but also because (as we have already seen in our discussion of emergence) the descriptive specification she chooses to make in her observation is *binding* with regard to how the "generative" processes — the relation between system and environment, system and element, organization and structure — can be construed. Once the observer has specified the system in question in her account of the phenomenon, the generative relations between organization and structure in the system being observed are *not* random or whimsical but must in fact be systematic. All of which is to say that the observation and explanation of a phenomenon constitute, de facto validate, and in this sense "generate" the relationship between the observed phenomenon (the "actual result of a process" of system plus environment) and the operations of the system that give rise to it. Most important, we must remind ourselves that the phenomenon and those generative operations take place in "nonintersecting domains" that become joined — but also potentially confused — in scientific explanation. As Maturana and Varela put it, "The problem begins when we unknowingly go from one realm to the other" — from the vantage of the environment to that of the system, both of which are joined by the observer in the observed "phenomenon to be explained" — "and demand that the correspondences we establish between them (because we see these two realms simultaneously) be in fact a part of the operation of the unity" (*Tree* 135–36). And this means, in turn, that we must attend assiduously to the distinction between operation and observation.

Necessary Blind Spots: Niklas Luhmann and the Observation of Observation

It is here that Niklas Luhmann's brilliant and innovative work on "the observation of observation" will prove invaluable. Luhmann's theory of observation attempts to make use of the much-maligned ocular metaphor by divorcing it from its representationalist associations, which are critiqued by Rortyan philosophy only to reappear in Rortyan politics. For Luhmann, all observations are constructed atop a constitutive distinction that is paradoxical or tautological, and that the observing system which utilizes the distinction cannot acknowledge *as* paradoxical and at the same time engage in self-reproduction. All systems, in other words, are constituted by a necessary "blind spot" that only *other* observing systems can see, and the process of social reproduction depends on the "unfolding," the distribution and circulation, of these constitutive paradoxes (which would otherwise block systemic self-reproduction) by a plurality of observing systems — not by observation but by "the observation of observation." Both Luhmann and Rorty begin from the Wittgensteinian position that "a system," as Luhmann puts it, "can see only what it can see. It cannot see what it cannot."[40] But Luhmann, unlike Rorty, derives from this formulation not the irrelevance of other observing systems (or Rortyan "beliefs") — not their exclusion from the conversation of social reproduction — but rather their very necessity.

Luhmann's theorization of the concept of observation and its relation to contingency is heavily indebted to the pioneering work of Maturana and Varela, but his refinement of the concept is a key component of his extension of the theory of autopoiesis from the realm of living systems (the focus of Maturana and Varela) to social systems as well. "If we abstract from life and define autopoiesis as a general form of system building using self-referential closure," Luhmann writes, "we would have to admit that there are nonliving autopoietic systems."[41] For Luhmann as for Maturana and Varela, the attraction of the concept of autopoiesis — or what Luhmann will more often treat under the term "self-reference" — is not least of all that the theorization of systems as both (operationally) closed and (structurally) open accounts for *both* high degrees of systemic autonomy *and* how systems change and "adapt" to their environments (or achieve "resonance" with them, as Luhmann puts it in *Ecological Communication*).[42]

But Luhmann extends and refines the work of Maturana and Varela in the particular theoretical pressure he applies to the problem of observation. It will come as no surprise that Luhmann agrees with Maturana and Varela that "Autopoietic systems . . . are sovereign with respect to the constitution of identities and differences. They, of course, do not create a material world of their own. They presuppose other levels of reality. . . . But whatever they use as identities and as differences is of their own making" (*Self-Reference* 3). But in essays like "Complexity and Meaning," Luhmann pushes beyond Maturana and Varela in his attention to the distinction between a system's *operation* and its *observation*. "By operation," he writes, "I mean the actual processing of the reproduction of the system." "By observation, on the other hand," he continues, "I mean the act of distinguishing for the creation of information" (*Self-Reference* 83). The distinction between operation and observation, Luhmann writes elsewhere, "occupies the place that had been taken up to this point by the unity-seeking logic of reflection. (*This means, therefore, a substitution of difference for unity*)" — about which we will say much more in a moment ("Cognitive Program" 68; my emphasis).

Luhmann distinguishes a third term here as well: *self-observation*. "Self-referential systems are able to observe themselves," he writes. "By using a fundamental distinction schema to delineate their self-identities, they can direct their own operations toward their self-identities" (*Self-Reference* 123). If they do not do so — if they cannot distinguish what is systemic and internal from what is environmental and external — then they cease to exist as autopoietic, self-producing systems. This is why Luhmann writes that the distinction between "internal" and "external" observation "is not needed," that "the concept of observation includes 'self-observation'" (*Self-Reference* 82). In other words, to observe *at all* requires an autopoietic system, and an autopoietic system capable of observation cannot exist without the capacity for self-observation — that is, without the capacity "to handle distinctions and process information."[43] Hence, observation and, within that, self-observation, are *themselves* necessary operations of autopoietic systems.

All of which leads us to the central point we need to understand about Luhmann's concept of observation and its relationship to the epistemological problem of constructivism. Luhmann's position is clearest, perhaps, in his explanation of the observation of observation in his important essay "The Cognitive Program of Constructivism and a Reality That Remains Unknown," where he writes: "An operation that uses distinctions in order to designate something we will call 'observation.' We are caught once again, therefore, in a circle: the distinction between operation and observation appears itself as an element of observation" (68–69). Most readers would probably agree with Luhmann — and beyond that, with the work of George Spencer Brown, which Luhmann draws on — that the most elementary cognitive act is to draw a distinction, to distinguish figure from ground, "x" from "not-x." The point Luhmann wishes to underscore, however, has been a familiar one ever since the "liar's paradox" of antiquity, or more recently, the theory of "logical types" of Russell and Whitehead, which tried to solve such antinomies:[44] that drawing such a distinction, the elementary constitutive act of observation, is always either paradoxical or tautological, and that this is both necessary and unavoidable. "Tautologies are distinctions," Luhmann writes,

> that do not distinguish. They explicitly negate that what they distinguish really makes a difference. Tautologies thus block observations. They are always based on a dual observation schema: something is what it is. This state-

ment, however, negates the posited duality and asserts an identity. Tautologies thus negate what makes them possible in the first place, and, therefore, the negation itself becomes meaningless. (*Self-Reference* 136)

To many readers, this description will evoke nothing so much as the famous Hegelian postulate of "the identity of identity and nonidentity." What Luhmann wishes to stress, however, is not the identity of identity and nonidentity but rather the *nonidentity* (or difference) of identity and nonidentity. As he puts it in *Ecological Communication,*

> the *unity* (of self-reference) that would be unacceptable in the form of a tautology (e.g. legal is legal) or a paradox (one does not have the legal right to maintain their legal right) is replaced by a *difference* (e.g. the difference of legal and illegal). Then the system can proceed according to this difference, oscillate within it and develop programs to regulate the ascription of the operations of the code's positions and counter-positions without raising the question of the code's unity. (xiv)[45]

Two points need to be stressed here. First, what enables this crucial emphasis on the *difference* of identity and nonidentity—it is also what separates Luhmann from the Kantianism with which he bears more than a passing affinity—is Luhmann's strident rejection of any possibility of a transcendental subject-observer. For Luhmann, all observations are produced by a *contingent* observer who could always, in theory, describe things otherwise. Hence, all observations—and all systems described by them—contain an irreducible element of complexity. As William Rasch puts it, for Luhmann—contra Hegel and Kant—"complexity can never be fully reduced to an underlying simplicity since simplicity, like complexity, is a construct of observation that could always be other than it is. Contingency, the ability to alter perspectives, acts as a reservoir of complexity within all simplicity."[46]

The second point that needs to be underscored in reference to Luhmann's position on tautology—and it is one whose pragmatic impulse will distinguish Luhmann's position from that of Derrida and deconstruction, at least in Luhmann's eyes[47]—is the insistence that the tautological (or, more strictly, paradoxical) nature of all observation constitutes a real, pragmatic *problem* for all social self-descriptions.[48] This is so, Luhmann argues, because "an observer can realize the self-referential systems are constituted in a paradoxical way. This insight itself, however, makes observation impossible, since it postulates an autopoietic system whose autopoiesis is blocked" (*Self-Reference* 139). The only way past this obstacle or blockage is that self-referential paradoxes must be—in Luhmann's somewhat frustrating nomenclature—"unfolded" by the system. We have already mentioned two ways in which such unfolding might take place: the unsatisfactory theory of logical types of Russell and Whitehead, which "interrupts" or unfolds the vicious circle of paradoxical self-reference "by an arbitrary fiat: the instruction to ignore operations that disobey the command to avoid paradoxes" (*EC* 24); and the operational reliance on binary coding, which enables the system to not so much "deparadoxize" itself as reorient its operations toward the *difference* of x and not-x (legal and nonlegal, for example) without ever raising the question of their paradoxical identity.

But if Luhmann's concern with the pragmatics of tautology and paradox for social reproduction separates him from Derrida and deconstruction, his position on how the practical-political "unfolding" of tautology and paradox ought to be handled separates him from consensus-seeking liberals such as Rorty or Habermas; for, if the processes of "deparadoxization" require that a system's constitutive paradox remain invisible to it, then the only way that this fact can be known as such is by an observation made by *another* observing system. As Luhmann puts it, "Only an [other] observer is able to realize what systems themselves are unable to realize" (*Self-Reference* 127). What is decisive about Luhmann's intervention here is his insistence on the constitutive blindness of all observations, a blindness that does not separate or alienate us from the world but, paradoxically, guarantees our connection with it. As Luhmann explains it in a remarkable passage:

> The source of a distinction's guaranteeing reality lies in its own operative unity [as, for example, legal versus not-legal]. It is, however, precisely as this unity [or paradoxical identity] that the distinction cannot be observed—except by means of another distinction which then assumes the function of a guarantor of reality. Another way of expressing this is to say the operation emerges simultaneously with the world which as a result remains cognitively unapproachable to the operation.
>
> The conclusion to be drawn from this is that the connection with the reality of the external world is established by the blind spot of the cognitive operation. *Reality is what one does not perceive when one perceives it.* ("Cognitive Program" 76; my emphasis)

Perception and cognition of reality, in other words, are made possible by the deployment of a paradoxical distinction to which the observation utilizing that distinction must remain "blind" if it is to perceive and cognize *at all*. Here, Luhmann neatly traverses what has traditionally seemed an insoluble epistemological problem: how to avoid the untenable reliance on the science/ideology distinction that has traditionally buttressed ideology critique and the sociology of knowledge, and at the same time avoid lapsing into epistemological solipsism. Luhmann's negotiation of this problem is possible *only* on the strength of systems theory's articulation of the observation of observation, which enables us to view the "blind spot" or "latency" of the observations of others not merely as ideological bias or the distortion of a pregiven reality knowable by "science," but rather as the unavoidably partial and paradoxical precondition of knowing as such.[49] This, Luhmann writes, is "the systematic keystone of epistemology—taking the place of its *a priori* foundation" ("Cognitive Program" 75). "In a somewhat Wittgensteinian formulation," he writes,

> one could say that a system can see only what it can see. It cannot see what it cannot. Moreover, it cannot see that it cannot see this....
>
> Nevertheless, a system that observes *other* systems has other possibilities.... [T]he observation of a system by another system—following Humberto Maturana we will call this "second-order observation"—can also observe the restrictions forced on the observed system by its own mode of operation.... It can observe the horizons of the observed system so that what they exclude becomes evident. (*EC* 23)

And here, we need to sharpen our sense of the pragmatic implications of Luhmann's epistemology and how it differs from Rortyan pragmatism. The passage we quoted earlier—that "the operation emerges simultaneously with the world which as a result remains cognitively unapproachable to the operation"—must surely remind us of Rorty's attempt to situate descriptions within a "nonintentional" and "causal" world *without* having either that world or the descriptions do representational work. But what follows in Luhmann—that "the connection with the reality of the external world is established by the blind spot of the cognitive operation," that "Reality is what one does not perceive when one perceives it"—separates Luhmann's crucial reformulation of the problem from Rorty. Luhmann stresses the contingency and paradoxicality of that very observation itself and—contra Rorty—derives from that contingency the *necessity* of the observations of others: it is only in the mutual observations of *different* observers that a critical view of any observed system can be formulated.

If we are stuck with constitutive distinctions that are paradoxical and must live with blind spots at the heart of our observations, Luhmann writes, "Perhaps, then, the problem can be distributed among a plurality of interlinked observers" who are of necessity joined to the world and to each other by their constitutive but different blind spots. The work of social theory would then consist in developing "thoughtful procedures for observing observation, with a special emphasis on that which, for the other, is a paradox and, therefore, cannot be observed by him" ("Sthenography" 137). And although this reformulation is neither, strictly speaking, a politics nor an ethics, it *does* provide a rigorous and persuasive theorization of the compelling necessity of *sociality as such*. It offers an epistemologically coherent and compelling model of necessary reciprocal and yet asymmetrical relations between self and other, observer and observed, relations that can no longer be characterized in terms of an identity principle (be it of class, race, or what have you) that would reduce the full complexity and contingency—the verticality, if you will—of the observer's position in the social space.

Politics, Ethics, and Systems Theory

In these terms, Luhmann's insistence on the "blind spot" of observation and, therefore, on the essential aporia of any authority that derives from it (the authority, say, of the system that enforces the distinction legal/illegal) bears more than a passing resemblance to the proposition of a fundamental "antagonism" at the core of social relations as theorized by Ernesto Laclau, Chantal Mouffe, and Slavoj Žižek. In a fuller discussion, we would want to pay particular attention, of course, to the differences between these theorists, and in particular to Žižek's rearticulation of social antagonism via Lacanian psychoanalysis.[50] Here, however, I want to focus on the tantalizing formal parallels between the theory of social antagonism and Luhmann's theory of the observation of observation, even if that form is finally *valued* differently by the theorists in question. As Žižek articulates the concept of social antagonism, "far from reducing all reality to a kind of language-game, the socio-symbolic field is conceived as structured around a certain traumatic impossibility, around a certain fissure which *cannot* be symbolized."[51] Or, to remind ourselves of Luhmann's formulation, "the connection with the reality of the external world is established by the blind spot of the cognitive operation. Reality is what one does not perceive when one perceives it." For Žižek as for Luhmann, "every identity is already in itself blocked, marked by an impossibility" (252), and thus "the stake of the entire process of subjectivation, of assuming different subject-positions"—or, in Luhmann's system, of a plurality of interlinked observers whereby paradox and tautology can be distributed in the social field—"is ultimately to enable us to avoid this traumatic experience" (253) (or Luhmannian operational "blockage") of the fact that it is our blind spot that assures our connection with the real.

For Žižek, the concept of social antagonism, which countenances "an ethics of confrontation with an impossible, traumatic kernel not covered by any ideal," constitutes "the only real answer to Habermas, to the project based on the ethics of the ideal of communication without constraint," because it unmasks the constitutive disavowal at work in models such as Habermas's: "I know very well that communication is broken and perverted, but still . . . (I believe and act as if the ideal speech situation is already realized)" (259). For Habermas, we will remember, complexity and contingency always contain the threat of relativism and even nihilism, and thus the proliferation of different systems of knowledge and value must be grounded in some sort of underlying simplicity. For Habermas—but not, significantly, for Rorty[52]—that simplicity is harbored in the very nature of language itself and its fundamental presupposition of an ideal speech act, of an undistorted communication through which the claims of different systems of thought and value can be adjudicated in a process of rational dialogue that arrives at common norms and values.[53] Žižek, however, like Luhmann, does *not* disavow the "broken and perverted" (i.e., paradoxical and tautological) nature of communication, but rather derives from that brokenness the necessity of sociality as such. He holds that "what this fetishistic logic of the ideal is masking, is, of course, the limitation proper to the symbolic field as such: the fact that the signifying field is always structured around a certain fundamental deadlock" (259) or what Luhmann characterizes as the "blockage" of paradoxical self-reference.

Like the theorists of social antagonism, then—and like them, against Habermas *and* against Rortyan ethnocentrism—Luhmann insists that the distribution of the problem of paradoxicality and the circulation of latent possibilities can take place only if we do *not* opt for the quintessentially modernist and Enlightenment strategy of the hoped-for *reduction* of complexity via social consensus. If all observation is made possible by a paradoxical distinction to which it must remain blind, then

> this is why all projection, or the setting of a goal, every formation of episodes necessitates recursive observation and why, furthermore, recursive observation makes possible not so much the elimination of paradoxes as their temporal and social distribution onto different operations. A consensual integration of systems of communication is, given such conditions, something that should sooner be feared than sought for. For such integration can only result in the paradoxes becoming invisible to all and remaining that way for an indefinite future. ("Cognitive Program" 75)

For Luhmann, the Habermasian strategy—or, for that matter, the Rortyan one of liberal recontainment of contingency via ethnocentrism—is a doomed and potentially dangerous project that might result in the blockage of communications and the "invisibilizing," rather than the unfolding and distribution, of paradox. Clearly, then, the Luhmannian concept of observation is not "intended to provide a ground-ing for knowledge, but only to keep open the possibility of observation operations' being carried out by very different empirical systems—living systems, systems of consciousness, systems of communications" ("Cognitive Program" 78). And just as clear too is Luhmann's resolute posthumanism, which concludes that what Haber-mas characterizes as the project of Enlightenment and modernity has—and must—come to an end. "With this," he writes,

> the traditional attribution of cognition to "man" has been done away with. It is clear here, if anywhere, that "constructivism" is a completely new theory of knowledge, a post-humanistic one. This is not intended maliciously but only to make clear that the concept "man" (in the singular!), as a designa-tion for the bearer and guarantor of the unity of knowledge, must be re-nounced. The reality of cognition is to be found in the current operations of the various autopoietic systems. ("Cognitive Program" 78)

There is a pragmatic premium in this philosophical difference, for in Luhmann's view the movement to a posthumanist perspective has the practi-cal benefit of enabling "better functional performance" (*EC* 128) of highly differen-tiated society and its component systems. For example, in *Ecological Communication*, Luhmann argues that "a sensible handling of system-theoretical analyses" will "lead more to the expansion of the perspectives of problems than to their suppression" (131). Such analysis, he contends, can provide an important counterbalance to de-structive social anxiety, which, for example, "is more likely to stop the effects of so-ciety on its environment, but...has to pay for this by risking unforeseeable internal reactions that again produce anxiety" (131)—and here, we might think of the "Spot-ted Owl controversy," where social anxiety about biodiversity and habitat destruc-tion did indeed "stop the effects of society on its environment," but at the expense of creating a severe generalized backlash of anxiety about environmental protection at the expense of economic well-being, one that threatened, ironically enough, to have severe repercussions for the reauthorization by Congress in 1993–94 of the Endangered Species Act, the very act that mandated the protection of the Spotted Owl in the first place!

It is important to note, however, that Luhmann makes it abun-dantly clear in many, many places that the pragmatic value of his theorization of complexity and functional differentiation is to enable *this world*—and, more specifi-cally, this liberal, Western, capitalist world—to engage in systemic self-reproduction without destructive blockages of autopoiesis, the better to achieve maximum resonance between the system and its environment.[54] Luhmann (and this is quite surprising, given his epistemological innovations) wholly takes for granted the enclosure of thought—even putatively revolutionary thought—by the Western liberal capitalist social system. As he puts it in *Political Theory in the Welfare State*, the basic problem for any would-be critical position is that

> every operational act, every structural process, every partial system participates in the society, and is society, but in none of these instances is it possible to discern the existence of the whole society. Even the criticisms of society must be carried out within society. Even the planning of society must be carried out within society. Even the description of society must be carried out within society.[55]

And although Luhmann would seem to register here nothing more than an epistemological truism, in fact he goes a good bit farther—as Danilo Zolo has pointed out—in his tacit endorsement of liberal capitalist society and "neoliberal" policies (a fact more than hinted at in Luhmann's political essays and in his systematically reductive glances at Marxist theory).[56] As Zolo puts it, Luhmann interprets

> the crisis of the welfare state in terms of the loss of the law's regulating ability. Accordingly, legislation invades private spheres as well as other functionally differentiated and autonomous sub-systems. In doing so, the welfare state's interventionist strategy overloads the law to the point of distorting its regulatory function. This overload results in chaotic legislation which complicates the legal system and prevents its rational self-reproduction. Against this, Luhmann and the reflexive law theorists defend the autopoietic autonomy of social sub-systems—particularly those concerning economy, education, and family life. Thus, the autopoietic paradigm supports deregulatory policies. (63)

To recall our discussion of Rorty, then, we might say that Luhmann, while he *does* evade pragmatism's "evasion of philosophy" and its reduction of complexity, *does not* evade a pervasive liberalism that, even more than in Rorty, takes the form of a technocratic functionalism that is content to operate wholly within the purview of what Lyotard has called the "performativity principle" of "positivist pragmatism."[57] In these terms, John McGowan's critique of Rorty would surely apply to Luhmann as well. As McGowan puts it, "the important thing to note is that the negative endorsement of change, of the ever continuing conversation"—or, we should add, of the continual unfolding of complexity and distribution of paradox in Luhmann's system—"is dependent upon and presupposes a much more positive version of the social world that the conversationalists inhabit."[58] As Richard Halpern has argued, this "quietism" is not necessarily an unavoidable consequence of Luhmann's work, and indeed he might be read in unexpected tandem with the post-Marxism of Laclau and Mouffe, Alec Nove, and others, which would help us see that "the movement from a capitalist to a post-capitalist society cannot be conceived of as a reduction of the complex to the simple, or the differentiated to the unified," but rather "would involve a movement towards greater social complexity, even hypercomplexity."[59] But Luhmann's complacent taking for granted of Western capitalist liberal society short-circuits one of the most politically promising aspects of his work: his rigorous theorization of the epistemological necessity and full complexity of sociality as such, of the fact that the social is always virtual, partial, and perspectival, mutually constituted by observers who can and must expose the aporias of one another's positions.

This shortcoming in Luhmann will be clearest, perhaps, if we compare Donna Haraway's reinterpretation of the figures of "observation" and vision in her recent work with that of Luhmann. Both Luhmann and Haraway attempt to retheorize the figure of vision by *situating* it—that is, by de-transcendentalizing it and divorcing it from its representationalist associations. Luhmann would agree, I think, with Haraway's insistence on "the embodied nature of all vision" and her rejection of "a conquering gaze from nowhere" that claims "the power to see and not be seen, to represent while escaping representation" ("Situated Knowledges" 188). And like Luhmann, Haraway's epistemological project is dedicated above all—to use her paraphrase of Althusser—to resisting "simplification in the last instance" (196). But here, Haraway's specific sense of "embodiment" as the name for this theoretical commitment needs to be distinguished from Luhmann's theorization of the contingency of all observation. What Haraway wants is a concept of "situated knowledges" (188) that emphasizes the physical and social positionality of the observer—not least of all, for Haraway, the observer's gender—the specific conjuncture of qualities that mark the possibilities and limits of what a specific observer can see. In Haraway's articulation of observation and vision, "embodiment" names contingency, "objectivity" names political and ethical responsibility for one's observations, and both are "as hostile to various forms of relativism as to the most explicitly totalizing versions of claims to scientific authority" (191).

There can be little doubt that Haraway would find in Luhmann's theorization of observation—his "unmarking" of it, we might say—confirmation of her suspicions about "relativism." Luhmann would need to be told, as Haraway reminds us, that "social constructivism cannot be allowed to decay into the radiant emanations of cynicism" ("Situated Knowledges" 184). Indeed, Luhmann would seem to invite this charge—both theoretically and tonally, rhetorically—in many places in his work. In *Ecological Communication*, for example, he writes:

> The problem seems to be that one has to recognize the dominant social structure—whether seen as "capitalism" or "functional differentiation"—to assume a position against it.... A functional equivalent for the [nineteenth-century] theoretical construct "dialectics/revolution" is not in sight and therefore it is not clear what function a critical self-observation of society within society could fulfill.... Like the "Reds"... the "Greens" will also lose color as soon as they assume office and find themselves confronted with all the red tape. (126)

My guess is that Haraway would detect—and would be justified in detecting—the leveling political extrapolation at the end of this passage from the epistemological claims at its beginning as an instance of that relativism which is, in her words, "a way of being nowhere and everywhere equally. The 'equality' of positioning is a denial of responsibility and critical enquiry. Relativism is the perfect mirror twin of totalization in the ideologies of objectivity; both deny the stakes in location, embodiment, and partial perspective" ("Situated Knowledges" 191).

Luhmann's theory of observation, then, does not sufficiently recognize the imperative of Haraway's "embodied objectivity": that "vision is *always* a question of the power to see" (192). Again, a passage from Luhmann's *Ecological Communication* will help to make the point:

> Investigations that are inspired theoretically can always be accused of a lack of "practical reference." They do not provide prescriptions for others to use.... This does not exclude the possibility that serviceable results can be attained in this way. But then the significance of theory will always remain that a more controlled method of creating ideas can increase the probability of more serviceable results—above all, that it can reduce the probability of creating useless excitement. (xviii)

But the question that is never broached by Luhmann is put squarely on the table by theorists like Haraway and Evelyn Fox Keller: "serviceable" *for whom?* And in the absence of addressing that question—and of any detectable *interest* in addressing it—Luhmann's position seems ripe for interpellation into Haraway's reading of systems theory in terms of the historically specific "management" strategies of post–World War II liberal capitalist society, in which systems theory (like sociobiology, population genetics, ergonomics, and other field models) is crucial to "the reproduction of capitalist social relations" in the specific era of "an engineering science of automated technological devices, in which the model of scientific intervention is technical and 'systematic,'...[t]he nature of analysis is technological functionalism, and ideological appeals are to alleviation of stress and other signs of human obsolescence."[60]

In Luhmann's hands, the systems theory paradigm finally does indulge in the same sort of blithe liberal functionalism embraced by Rorty in its refusal to confront the uneven and asymmetrical relations of power—especially economic power—that undeniably constrain and indeed often render utterly beside the point the unfolding of complexity and the distribution of paradox which remain in Luhmann's thought too squarely within a political if not philosophical idealism. If Rorty sanitizes the social field by limiting conversation to the liberal *ethnos*, Luhmann levels it by refusing to complicate his epistemological pluralism—that we are all alike in the formal homology of our observational differences—with an account of how in the material, social world in which those observations take place some observers enjoy more resources of observation than others. The complexifying and open-ended imperative of Luhmann's theory is, following George Spencer Brown, "distinguish!" and "observe!" but we must still subject that imperative to the critique leveled by Steven Best and Douglas Kellner at the metaphor of cultural "conversation" of diversity and plurality as it is deployed by Rorty: "that some people and groups are in far better positions—politically, economically, and psychologically—to speak [or to observe, we might add] than others. Such calls are vapid when the field of discourse is controlled and monopolized by the dominant economic and political powers" (*Postmodern Theory* 288).

We might say, then—to broach a topic I will take up in my conclusion—that *Luhmann's* "blind spot," *his* unobservable constitutive distinction, is his unspoken distinction between "differentiation" and what historicist, materialist critique has theorized as "contradiction," a blind spot that manifests itself in Luhmann's inability or unwillingness to adequately theorize the discrepancy between the formal equivalence of observers in his epistemology and their real lack of equivalence on the material, social plane. It seems that the category of contradiction—insofar as it names precisely this difference—proves more difficult to dispose of than Luhmann's systems theory imagines. Or rather—to put a somewhat finer point on it—it is disposed of by systems theory, but only "abstractly," as Marxists theorists like to say, only in thought, but not in historical, material practice.

To put the issue in this way is to raise perhaps the most far-reaching question of all: whether or not postmodern society is adequately described as primarily functionally differentiated, as in Luhmann's account, or stratified and hierarchical, as in Marxism's continued insistence on the priority of the economic. In this connection, we might (to stay with the Marxist critique for a moment) invoke Raymond Williams's famous revision of the base/superstructure model to say that functional *differentiation* is *emergent*—even though it might be more pervasive socially—within a system in which dialectical *contradiction* remains *dominant* in the form of the asymmetrical importance of the economic system.[61] In that light, what Luhmann's epistemological idealism refuses to confront is that the differentiation, autonomy, and unfolding of complexity it imagines remains muffled and mastered by the economic context of identity and exchange value within which systems theory itself historically arises. And in that refusal, in its pragmatic effect of socially reproducing the liberal status quo, it is clear that there are powerful ideological reasons, as well as epistemological ones, why one cannot see what one cannot see.

This does not mean, however, that systems theory *tout court* is subject to the sorts of political critiques most commonly leveled at it by philosophers and historians of science such as Haraway and Peter Galison, ecological feminists such as Carolyn Merchant, and popular social critics such as Jeremy Rifkin.[62] Merchant's critique is standard: that the systems theory paradigm can "be appropriated, not as a source of cultural transformation, but as an instrument for technocratic management of society and nature, leaving the prevailing social and economic order unchanged" (104). It is true, as Steve J. Heims points out in his social history of the Macy cybernetics conferences of 1946–53 (*Constructing a Social Science for Postwar America*), that the conferences themselves were conducted in the stringently apolitical atmosphere of the Cold War that hung over first-order cybernetics as a whole, an atmosphere in which questions of politics, ideological differences, and alternative social configurations were strongly discouraged, if not forbidden. But if these sorts of critiques may be valid for first-order cybernetics, it is difficult to see how they would hold for second-order cybernetics, with its emphasis on the radical contingency of observation, the embodiment of knowledge, and the irreducible complexity of systemic description that flows from both. As we have already seen, second-order cybernetics, by pursuing the full implications of the principle of recursivity held at bay in its predecessor, concerns itself at least as much with the creative, emergent, and unpredictable capacities of self-organizing and autopoietic systems as with the mechanisms of control and closure foregrounded by the Macy conferences. And although second-order systems theory does make a claim to universal descriptive veracity, that claim is mounted on its ability to theorize the *inability* to see the social or natural system as a totality from any particular observer's point of view. It is difficult, therefore, to see how second-order cybernetics could justly be described as in principle a theoretical instrument of globalized "technocratic management" when it foregrounds the very contingency, complexity, and unpredictability that such programs of technocratic control would want to suppress.

It is more useful, I think—and more a propos the theoretical commitments of second-order cybernetics—to reframe the work of systems theory (despite its shortcomings in Luhmann's hands) in terms of what Merchant calls the need for "reconstructive knowledge" that should be based on "principles of interaction (not dominance), change and process (rather than unchanging universal principles), complexity (rather than simple assumptions), contextuality (rather than context-free laws and theories), and the interconnectedness of humanity with the rest of nature" (107). If it seems far-fetched to read the second-order cybernetics of Maturana and Varela in this light, we should remember that they themselves have cast the pragmatic and ethical import of their theoretical work very much in these terms. As they put it at the end of *The Tree of Knowledge:*

> The *knowledge of knowledge compels.* It compels us to adopt an attitude of permanent vigilance against the temptation of certainty. . . . It compels us to realize that the world everyone sees is not *the* world but *a* world which we bring forth with others. It compels us to see that the world will be different only if we live differently. (245)

Maturana and Varela understand, as does feminist philosophy of science in its own way, that the stakes over the epistemological status of "objectivity" are far from purely epistemological. But Maturana and Varela base the ethical and pragmatic value of their work squarely on the difference between the epistemology of representationalism and realism ("knowledge") retained by feminist philosophy of science, and they set out from the second-order theorization of the problematics of contingent observation, from the fact that "everything said is said by someone" ("knowledge of knowledge"): "*We affirm that at the core of all the troubles we face today is our very ignorance of knowing.* It is not knowledge, but the knowledge of knowledge, that compels" (248). The "knowledge of knowledge" leads Maturana and Varela to now conclude, in a quite remarkable passage, that second-order cybernetics "implies an ethics we cannot evade":

> If we know that our world is necessarily the world we bring forth with others, every time we are in conflict with another human being *with whom we want to remain in co-existence,* we cannot affirm what for us is certain (an absolute truth) because that would negate the other person. If we want to co-

> exist with the other person, we must see that *his certainty — however undesirable it may seem to us — is as legitimate and valid as our own.* . . . Let us not deceive ourselves; we are not moralizing, we are not preaching love. We are only revealing the fact that, biologically, without love, without acceptance of others, there is no social phenomenon. (246–47)

It is hard to imagine a more powerful statement of the ethical imperatives of second-order cybernetics than this.

Unfortunately, it is also hard to imagine a more powerful symptom of the unreconstructed humanism that is just as inadequate to the epistemological innovations of second-order cybernetics as the "objectivist" epistemology of feminist philosophy of science is to its progressive political agenda. That humanism manifests itself in Maturana and Varela in the philosophical idealism that hopes that ethics may somehow do the work of politics. What we find here, in other words, is (to borrow Fredric Jameson's formulation) a kind of "strategy of containment" whereby the posthumanist imperatives of second-order cybernetics are ideologically recontained by an idealist faith in the social and political power of reason, reflection, voluntarism, and what Jameson calls "the taking of thought"[63]: "*We affirm that at the core of all the troubles we face today is our very ignorance of knowing.*"

My point is not to take issue with Maturana and Varela's emphasis on the importance of "bringing forth a common world," but rather to remind them, as Jameson puts it in *The Political Unconscious*, that ethical thought "projects as permanent features of human 'experience,' and thus as a kind of 'wisdom' about personal life and interpersonal relations, what are in reality the historical and institutional specifics of a determinate type of group solidarity or class cohesion" (59). It is precisely this contradiction that lies buried in Maturana and Varela's crucial but subordinated proviso, "*with whom we want to remain in co-existence.*" Maturana and Varela's ethical assertion of the necessity of love is predicated on the assumption that the problem of social (including economic and class) difference that Jameson highlights has always already been solved. In the process, Maturana and Varela drain the assertion of contingency of its materialist, pragmatic force, whose entire point — as we know from feminist philosophy of science as well as Marxist theory — is to say that all points of view are *not* equally valid precisely because they have material *effects* whose benefits and drawbacks are distributed asymmetrically in the social field. And this asymmetry, in turn, makes it vastly easier for some groups and persons to enjoy the luxury of freely accepting the "validity" of points of view other than their own. This, after all, is the point of Fox Keller's assertion that the practice of knowledge always works *at* something specific and *for* a particular "we." Maturana and

Varela are right that, epistemologically speaking, all points of view are equally contingent; but this does not mean, from a pragmatic point of view, that we need treat all points of view as equally "legitimate and valid." Indeed, as Ashmore et al. point out, "if objective truth and validity are renounced in favor of social process and practical reasoning, then so also must be any notion of a commitment to '*equal* validity.' Far from ruling out the possibility of justification of a particular view, relativism insists upon it" (10).

Such advice seems even more crucial to remember in light of the use to which Buddhist philosophy is put in Varela, Thompson, and Rosch's *The Embodied Mind*. In chapter 10 of that study, for example, Varela et al. want to distinguish their Buddhist commitment from Western pragmatism proper, and they argue that "Western philosophy has been more concerned with the rational understanding of life and mind than with the relevance of a pragmatic method for transforming human experience" (218). But what becomes clear in later chapters is that this "pragmatic method" consists of repeated calls for us to heed the wisdom of Buddhist "mindfulness" and "egolessness" to solve by ethical fiat and spiritual bootstrapping the complex problems of social life conducted in conditions of material scarcity, economic inequality, and institutionalized discrimination of various forms. This is especially clear in their critique of Garret Hardin's "The Tragedy of the Commons," where they respond to the problem of scarcity and the self-interested conduct it generates in terms already familiar from *The Tree of Knowledge*: "We believe that the view of the self as an economic man, which is the view the social sciences hold, is quite consonant with the unexamined view of our own motivation as ordinary, non-mindful people" (246). And the "pragmatic" answer to self-interested conduct created by conditions of economic scarcity, they tell us, is *not* to address that material scarcity and inequality itself, but rather to encourage through enlightenment "an attitude of all encompassing, decentered, responsive, compassionate concern," which "must be developed and embodied through a discipline that facilitates letting go of ego-centered habits and enables compassion to become spontaneous and self-sustaining" (252).

But clearly, as we have already suggested, this amounts to little more than telling people that the problems of scarcity and the maldistribution of wealth and power will stop being problems if we all simply stop being so selfish — a claim, of course, that is very easy for some to make and very hard for some to hear. Here, as elsewhere in Maturana and Varela, the complicated relationship between ethics and politics is not so much explained as explained away by an appeal to total human transformation with little or no attention to the material factors that make such an appeal little more than wishful thinking. And from this vantage, "love" as Maturana and Varela define it can in fact be *anti*social, *even if* it preserves "the biologic process that generates" the social process. In the end, then, Maturana, Varela, et al. give us "embodiment," but not a robust, *socially and historically situated* embodiment, and their "pragmatism" is disabled by exactly what is criticized in Husserl in *The Embodied Mind:* that the "self" and its "experience"—the linchpins of their critique of formalist epistemology—remain "entirely *theoretical*" and lack any "*pragmatic* dimension" (*Embodied Mind* 19). As Vincent Kenny and Philip Boxer put it in their comparison of Maturana and Lacan, "What *does* make the difference between the family, the asylum and the concentration camp as forms of social structural coupling? If there are those who would argue that these are all the fruits of reflection and an 'opening up of room for existence,' are reflection and love enough therefore as an ethics?"[64]

The answer would seem to be "no," not only for Jamesonian reasons but also, as it were, for post-Jamesonian ones: that Maturana and Varela's call for an ethic of love constitutes a radical disavowal of the fundamental social antagonism that we have already examined, one whose form Žižek characterizes as: "I know very well there are views which I despise, but still..." ("Beyond Discourse-Analysis" 259).[65] What Maturana and Varela disavow is nothing other than the "auto-negativity" and "self-hindering" status of the subject and its desire, its lack, its traumatic "internal limit"; indeed, "the stake of the entire process of subjectivation, of assuming different subject-positions," Žižek writes, "is ultimately to enable us to avoid this traumatic experience" (253). As Žižek puts it, "'the subject' in the Lacanian sense is the name for this internal limit, this internal impossibility of the Other, of the 'substance.' The subject is a paradoxical entity which is so to speak its own negative, i.e. which persists only insofar as its full realization is blocked—the fully realized subject would be no longer subject but substance" (254). We will remember that the Lacanian name for this substance is, of course, the Real, or what Kant, in the *Critique of Practical Reason*, called the "pathological" Thing, *das Ding*. And in this light, it becomes clear that Maturana and Varela's terrifying injunction ("Love!") is, from a psychoanalytic point of view, a call for an end to the problem of desire, a call for the continued repression of the Thing at the heart of the subject—of the "biology," if you will, at the heart of the "biological process."

When we recall, moreover, that the most familiar name for substance, *das Ding*, and the Real since Freud's *Civilization and Its Discontents* is the *animal*, then what moves strikingly and quite surprisingly into view is that the surest sign of Maturana and Varela's persistent humanism is not their individualism, nor even their idealism, but rather the systematic *speciesism* that is unmistakable in their work separately and in collaboration. It is not simply that Maturana and Varela frame their ethics solely in terms of the reciprocal relations between *human beings*, and in doing so undercut the promise of their epistemology by leaving aside the very posthumanist imperatives—of ecology, of animal rights, of the political and ethical challenges of technoscience—which we mentioned at the beginning. It is rather the sort of jarring, symptomatic contradiction on which their ethical project runs aground again and again: on the one hand, it persuasively argues (following groundbreaking work in cognitive ethology over the past two decades) that the human species is not the only one to participate in social, cultural, and linguistic domains, and it recognizes the importance of individual temperament and ontogeny for social organization and communication among nonhuman animals—all of which are factors that, by their own definition, constitute grounds for ethical consideration.[66] On the other hand, their work systematically invokes and praises some of the most invasive and brutal animal research on monkeys, cats, rabbits, and other nonhuman animals conducted in recent decades.

This quintessentially humanist "blind spot" constitutes an almost unbearable myopia in Varela et al.'s *The Embodied Mind*, where the authors call for "the cultivation of compassion for all sentient beings" (248), for a "responsiveness to oneself and others as sentient beings without ego-selves" (251). And then, having issued such a call, they proceed to praise the extremely controversial neurophysiological research of Russell DeValois on macaque monkeys (170ff.) (which has been challenged for a decade for its brutality and frivolity by several leading animal rights groups), and recount a "beautiful study" in which kittens were raised in the dark, kept entirely passive, and as a result when released "after a few weeks of this treatment" acted "as if they were blind: they bumped into objects and fell over edges" (175).

This blindness on the part of the authors, however, will perhaps come as less of a surprise when we psychoanalyze it, when we remember that the relationship between subject and substance in the Enlightenment paradigm as articulated by Žižek is one of traumatic disavowal of the bond between meaning and substance, self and thing, human and animal. In this light, the surest sign of humanism is that "subjectivation designs the movement through which the subject integrates what is given him/her into the universe of meaning—[but] this integration always ultimately fails, there is a certain left-over which cannot be integrated into the symbolic universe, an object which resists subjectivation, and the subject is precisely the correlative to this object" ("Beyond Discourse-Analysis" 254). Maturana and Varela hope that "love" will achieve such an integration, but it is clear that the most quintessentially humanist "leftover" in their discourse, as in humanism generally, is the *animal other* as articulated by the discourse of speciesism, with the subject of humanism its precise correlative. Maturana and Varela's humanist ethics thus fails precisely because it *is* humanist; it attempts to solve by ethical fiat the posthumanist political challenges that their epistemology, as a possible "reconstructive form of knowledge," might help us to theorize. Their ethics forgets what their epistemology knows: that in the cyborg cultural context of OncoMouse™ and hybrids of nature/culture, the question is not who will get to be human, but what kinds of couplings across the humanist divide are possible — or unavoidable — when we begin to observe the end of Man. A frontal engagement with those very questions will occupy the subjects of our next chapter, Michel Foucault and Gilles Deleuze.

THREE

Poststructuralism

Foucault with Deleuze

THIS CHAPTER sets out from the convergence of Michel Foucault and Gilles Deleuze on the problem that lends its name to this study, the problem of the "outside." To delimit in this way our examination of the huge body of work produced by both is to at the same time tighten the focus on the specific brand of poststructuralism that joins Foucault and Deleuze and separates them from other theorists often treated under the same rubric: their engagement not only with what Deleuze will characterize as "the form of expression" but also with "the form of content." For both Foucault and Deleuze, "one's point of reference," as Foucault puts it, "should not be to the great model of language [*langue*] and signs" because the "history which bears and determines us" does not have the form of a language. For both, then, the theoretical challenge at hand is always "relations of power"—or of "force" in Deleuze's lexicon—"not relations of meaning."[1] My reading thus takes for granted the view expressed by one of our best readers of Deleuze, Brian Massumi, who argues that what separates Deleuze (and Deleuze's Foucault) from other putatively poststructuralist thinkers—and more specifically from deconstruction—is that the latter "does not allow for the possibility of a positive...description of nonbinary modes of differentiation. It leaves the identity-undifferentiation system basically intact, emphasizing the ineffability, unthinkability, and unsustainability of what subtends identity."[2] On this view, many poststructuralists continue "to repose in the shadow of Saussure's tree" (178 n. 73); they continue, that is, an essentially structuralist, diacritical project. For Deleuze, on the other hand, "the singular, the 'heteroclite,' is not 'confused' and unanalyzable. It simply obeys other, far more complex, rules of formation" (91). It is this commitment to theorizing the positivity of the singular, the heteroclite, the contingent, the form of content, that distinguishes the poststructuralism of Deleuze and Foucault as an ambitious if often vexed engagement with the "outside" of theory.

Rorty and Foucault

The Foucault who emerges on the terrain of the "outside"—Deleuze's Foucault, if you will—is quite a different figure from the Foucault presented to us in Richard Rorty's reading. It would be possible, of course, to fruitfully examine Foucault in light of the other pragmatist figures I have discussed thus far. Walter Benn Michaels, like many New Historicists, explicitly draws on Foucault's critique of ideology (in *The Gold Standard and the Logic of Naturalism*), and the figure of Foucault looms large in Frank Lentricchia's *After the New Criticism* and *Ariel and the Police*. As for Stanley Cavell, his picture of philosophy as a task of "onwardness," as the "aversion of conformity," would bear detailed comparison with Foucault's view of philosophical thought as a dynamic and transgressive "force of flight," always seeking a "limit experience."[3] But I want to dwell on the comparison with Rorty, not least of all because Rorty has explicitly engaged Foucault's work in a series writings, from a 1979 lecture included in David Couzens Hoy's *Foucault: A Critical Reader*, through part of a chapter in *Consequences of Pragmatism*, and into remarks on Foucault in the later books *Contingency, Irony, and Solidarity*, *Objectivity, Relativism, and Truth*, and *Essays on Heidegger and Others*.

The general similarities between Rorty and Foucault are easy enough to sketch. Both are committed to viewing the individual as a culturally constructed "web of beliefs" (Rorty) or a product of social practices of subjection (Foucault) rather than as a natural or transhistorical entity. Both critique the assumption that "Reason" is a transcendental pursuit, and hence both rightly have been called antihumanists, if by "humanism" we mean the view that the role of culture and philosophy is to bring to light and develop a given, inner human nature. The similarities between the two are perhaps best summed up by Rorty himself, whose version of Foucault we touched on earlier. In "Moral Identity and Private Autonomy: The Case of Foucault," Rorty follows Vincent Descombes's suggestion that there are *two* Foucaults. The first is an "American" one who, Descombes writes, "sought to define autonomy in purely human terms," and so can be read in Rorty's eyes as an "up-to-date version" of John Dewey.[4] Like Dewey, Rorty's American Foucault "tells us that liberal democracies might work better if they stopped trying to give universalistic self-justifications, stopped appealing to notions like 'rationality' and 'human nature' and instead viewed themselves simply as promising social experiments" (193). But, as Rorty notes in his development of the comparison in the earlier study *Consequences of Pragmatism*, once you have adopted the pragmatist position, there are still two directions in which you can take it. "Dewey," Rorty writes, "emphasizes that this move 'beyond method' gives mankind an opportunity to grow up, to be free to make itself, rather than seeking direction from some imagined outside source," whereas Foucault "views this move as the Nietzschean realization that all knowledge-claims are moves in a power-game."[5]

It is precisely at that juncture that for Rorty we pass over into the "other" Foucault, into what Descombes calls "the French Foucault," the *"fully Nietzschean"* one who is not merely the antifoundationalist critic of transcendental philosophy (like Rorty), but who (also like Rorty) also urges us to engage in ceaseless self-invention, endless rearticulation, and to "have thoughts which no human being has yet had" (*EHO* 193). This Foucault, in other words, is what Rorty calls an "ironist," a variety of "Romantic intellectual" who sees himself as a "knight of autonomy" (194). For Rorty, this Nietzscheanism in Foucault would pose no problem were it not for the fact that Foucault often wants to extend his desire for autonomy into the public realm, and thus sometimes violates Rorty's "partition" between public and private that we have already discussed. The problem with Foucault's Nietzscheanism, in other words, is that Foucault is not willing to keep it to himself. "It is only when a Romantic intellectual begins to want his private self to serve as a model for other human beings," Rorty writes,

> that his politics tend to become antiliberal. When he begins to think that other human beings have a moral duty to achieve the same inner autonomy as he himself has achieved, then he begins to think about political and social changes which will help them to do so. Then he may begin to think that he has a moral duty to bring about these changes, whether his fellow citizens want them or not. (194)

For this reason, Rorty concludes, "Insofar as the French Foucault has any politics, they are anarchist rather than liberal" (193).

And that politics comes in for harsh criticism indeed from Rorty (and from those he quotes approvingly) as "rhetoric and posturing" (194), as "self-indulgent radical chic," "anarchist claptrap about repression," and "Nietzschean bravura about the will-to-power."[6] We should probably be suspicious on principle toward charges of "claptrap about repression" from one who is all too ready to declare, as Rorty is, that "there is nothing wrong with liberal democracy."[7] And it also worth remembering, as one recent critic points out, that Foucault's lack of concern with meeting the approval of the "established regimes of thought" makes him "a *bête noire* of mainstream or liberal political theorists" such as Habermas, Fraser, Walzer, Taylor—and, of course, Rorty himself.[8] But while charges such as Rorty's are no doubt shrill, they do contain, I think, an element of truth. Foucault is indeed often, as Charles Taylor has argued, amazingly "one-sided" in his wholesale critique of the Enlightenment and modernity (Rorty, *EHO* 195); for example, his reading of modernity ignores, as Habermas puts it, how the "eroticization and internalization of subjective nature also meant a gain in freedom and expression."[9]

But if Foucault's work is often one-sided, I now want to argue that it is a one-sidedness motivated by a critical pragmatism that is designed, in its emphases on power and its critique of the idea of Man, to alert us to precisely the sorts of problems we find in Rortyan pragmatism. Foucault's critique, overstated though it may sometimes be, would force us to see, first, that the "truth" of Rorty's liberal pragmatism is not to be found so much in its philosophical antirepresentationalism as in the social and material forms of its realization; and second, that Rorty's attempt to keep the public and private separate is—in light of Foucault's *The History of Madness, Discipline and Punish*, and *The History of Sexuality*—not only untenable but indeed constitutes a crucial mechanism for masking the operation of social power and its forms of subjectification. As Honi Haber points out, Rorty's private/public partition must be rejected because it "refuses to see ideas of beauty, of desire, of normalcy, of intelligence, as being constituted within a public/political discourse and so subject to instrumental, or 'normalizing' and 'disciplinary' rationality" (75–76).

On a different level of analysis, Rorty's partitioning of the "good" and "bad" Foucault—and his transformation of the "American" Foucault into an erstwhile John Dewey—also operates in the services of an American exceptionalism that, as Tom Cohen has noted, defines its liberal Self, its "inside," against the Other, the "outside," of continental (read: "bad" Nietzschean and "anarchist") theory.[10] This sort of intellectual provincialism in Rorty makes him unable to view Foucault within the very context needed to provide an antidote to Rorty's most jarring misreading—or rather "underreading"—of Foucault, and that is Foucault's contribution to post-Marxist critique (an issue to which we will return). This should come as no surprise, however, because Rorty has shown himself consistently incapable of serious engagement with Marxist theory, a failure that usually clothes itself in Rorty's thread-bare characterization of Marxism in terms of redemptionist and more or less theo-logical schemes.[11] This characterization from *Consequences of Pragmatism* is typical: "Man as Hegel thought of him, as the incarnation of the Idea, doubtless does have to go. The proletariat as the Redeemed Form of Man has to go, too. But there seems no particular reason why, after dumping Marx, we have to keep on repeating all the nasty things about bourgeois liberalism which he taught us to say" (207). Rorty's American exceptionalist inability to view Foucault in light of the Marxist tradition leads him to a fundamental misreading of the sort that we find at the end of that same essay, where he writes that "Foucault's vision of discourse as a network of power-relations isn't very different from Dewey's vision of it as instrumental, as one element in the arsenal of tools people use for gratifying, synthesizing, and har-monizing their desires" (208).

Leaving aside whether or not this is an apt characterization of Dewey, it is clear that what Rorty misses here is the very core of Foucault's work on the relationship between knowledge and power. The picture Rorty gives us of an undifferentiated (liberal pluralist) "people" taking thought, "synthesizing" and "har-monizing" their desires, and then reaching out for the appropriate tool to cash in those desires, could not be more unlike the picture of power/knowledge that Fou-cault gives us. As we have already noted, it ignores the relationship between those desires (supposedly "private" in Rorty's reading) and the sites of their social produc-tion. For, as Foucault notes in one of the most important passages in *Discipline and Punish*,

> Historically, the process by which the bourgeoisie became in the course of the eighteenth century the politically dominant class was masked by the es-tablishment of an explicit, coded and formally egalitarian juridical frame-work, made possible by the organization of a parliamentary, representative regime.... And although, in a formal way, the representative regime makes it possible, directly or indirectly, with or without relays, for the will of all to form the fundamental authority of sovereignty, the disciplines provide, at the base, a guarantee of the submission of forces and bodies. The real, cor-poreal disciplines constituted the foundation of the formal, juridical liber-ties.... The "Enlightenment," which discovered the liberties, also invented the disciplines....
>
> Regular and institutional as it may be, the discipline, in its mechanism, is a "counter-law." And, although the universal juridicism of modern society seems to fix limits on the exercise of power, its universally widespread panop-

ticism enables it to operate, on the underside of the law, a machinery that is both immense and minute, which supports, reinforces, multiplies the asymmetry of power and undermines the limits that are traced around the law.[12]

In contrast to Rortyan "belief"—which begins to look in this light like the concept of "ideology" that Foucault rejects because it presupposes "something of the order of a subject" ("Truth and Power" 60)—Foucault's rendering of the relationship between power and knowledge through the example of the disciplines as a kind of "counter-law" aims to emphasize the binding, subjectivizing *materiality* of practice. The relationship between the subject with his desires and the proper tools for implementing those desires is neither, as Rorty's reading of Dewey suggests, fully masterable by the subject, nor is it unidirectional; the process has an *unconscious*, in other words. Rorty's pragmatism paints an oversimplified picture of the malleability of those "tools" that one chooses to use, and it essentially ignores the countervailing, subjectivizing force of the "tools" that use the subject just as surely as the subject uses them. Moreover, as one of our most helpful readers of Foucault, Barry Smart, has pointed out, "it is the problematic or uneven character of the relations between such rational schemas or programmes, associated social and institutional practices and their 'unintended' or 'unprojected' effects that has constituted the focus of much of Foucault's work."[13] For these and other reasons, Foucault cannot be enlisted into the Rortyan "conversation" model of social interaction.[14] Rorty's characterization of the subject as a "web of beliefs" *seems* to recognize that the subject who speaks is also always already the subject who is spoken, who says more than and *other* than she intends. But Rorty's partition position takes away with one hand what his concept of belief seems to give with the other by rendering the "tools" used by the Deweyan subject as pure means that come into play only at the *end* of the process of social production, and not as formative—and we might say, "counterformative"—of the Deweyan subject's "private" desires that *initiate* the process.

What this means, then, is that Rorty's reading of Foucault fails to understand the importance of Foucault's theorization of power's *productive* (rather than strictly repressive) aspect, a failure that shows up in the untenability of Rorty's use of the avoidance of pain as the criterion for adjudicating disputes between private irony and public liberalism.[15] As Smart points out:

> The historical transition documented [in *Discipline and Punish*] is conceived by Foucault not in terms of a reduction in the use of "violence" in the exercise of power and a concomitant increase in "consent," but in terms

> of . . . a new form of "pastoral power" over the social, that is to say the deployment of various measures directed to the health, well-being, security, protection, and the development of both the individual and the population . . . the development of individualizing techniques and practices which are reducible neither to force nor to consent, techniques and practices which have transformed political conflict and struggle through the constitution of new forms of social cohesion. (161–62)

From this vantage, we can see that Rorty's position misses the entire point of the critique of ideology from Marx and Engels to Žižek, Laclau, and Mouffe—a tradition to which Foucault, despite his attempts to distance himself from that lineage, may be seen as a key contributor.[16]

It is well known, of course, that Foucault roundly rejected the concept of ideology for reasons which, in retrospect, seem to have as much to do with distinguishing himself from the Marxism of his former teacher Louis Althusser as with embracing the latter's more structuralist, "antihumanist" elements.[17] For Foucault, the concept of ideology, as we have noted, presupposes a "constituent subject" and remains too firmly tied to the base/superstructure model of traditional Marxism in which "superstructural" elements (such as Foucault's "disciplines" and "practices") are thought to be determined by the economic—a position that Foucault explicitly rejects in *Discipline and Punish* and elsewhere.[18] Nevertheless, Foucault's work may be seen as a crucial contribution to the theorization of the *materiality* of ideology pursued by some of the most important Marxist and post-Marxist thinkers from Althusser through Žižek. The force of Foucault's work on the microdisciplines and the techniques of subjection and normalization is to remind us that the new forms of social cohesion that he studies depend less on what the subject *thinks* than what the subject *does*. In contrast to the humanist and more explicitly phenomenological Marxism of Lukács and Gramsci, where critical consciousness is a central concern,[19] Foucault's point about the techniques of discipline is that the new modes of social cohesion will be content if you are performing your marching techniques with discipline, *even if* you are thinking of Keats's "Ode on a Grecian Urn" (or even Marx's *Eighteenth Brumaire*) while you are doing so. Indeed, an even stronger reading of Foucault would say that the humanist belief that critical consciousness will set you free is itself an essential ruse of the system that, in the name of "well-being" and "individual development," channels revolutionary desire into merely *thinking* freedom rather than acting it. Foucault's point is not so much that, in the new modes of social cohesion, it does not matter what you think, but rather that society is willing to give the subject generous latitude in this regard—is even willing to nurture the subject and provide therapeutic and educational support to this end—so long as the subject undergoes the disciplines of subjectification that distribute her in an analytic social space, render her visible and knowable, and lead her to inscribe the mechanisms of discipline on her very body, in her very actions, through regimes of "health," sexuality, and other "technologies of the self."[20]

It is this crucial dimension of Foucault's work on the relationship between power and knowledge that seems totally missed by Rorty, and that links Foucault with the development of the concept of ideology in Althusser's influential essay on ideological state apparatuses, where he clarifies his thesis that "ideology has a material existence" by reference to a well-known example from Pascal "which will enable us," he writes, "to invert the order of the notional schema of ideology. Pascal says, more or less: 'Kneel down, move your lips in prayer, and you will believe.' "[21] Slavoj Žižek glosses this crucial moment in the development of ideology theory, which reorients us toward viewing "ideology in its otherness-externalization":

> Religious belief, for example, is not merely or even primarily an inner conviction, but the Church as an institution and its rituals (prayer, baptism, confirmation, confession...) which, far from being a mere secondary externalization of the inner belief, stand for *the very mechanisms that generate it*.... That is to say, the implicit logic of his argument is: kneel down and *you shall believe that you knelt down because of your belief*—that is, your following the ritual is an expression/effect of your inner belief, in short the "external" ritual performatively generates its own ideological foundation.[22]

It is this material aspect of ideology that Žižek has developed in his own complicated and compelling work on "the reality of the ideological fantasy" and what he calls "the objective status of belief"—which we would do better to read, I think as "object*al* status."[23] As Žižek puts it, "it is belief which is radically exterior, embodied in the practical, effective procedure of the people. It is similar to the Tibetan prayer wheels," he continues in an amusing and instructive example:

> You write a prayer on a paper, put the rolled paper into a wheel, and turn it automatically, without thinking.... In this way, the wheel itself is praying for me, instead of me—or, more precisely, I myself am praying through the medium of the wheel. The beauty of it all is that in my psychological interiority I can think about whatever I want, I can yield to the most dirty and obscene fantasies, and it does not matter because—to use a good old Stalinist expression—whatever I am thinking, *objectively* I am praying. (*Sublime Object* 34)[24]

In this light, if there is a reason to retain the notion of ideology in the face of proposals by Michaels, Rorty, and others that we abandon it for "belief," it is *not* because of some untenable science/ideology distinction that would enable us to condemn some notions as examples of "false consciousness." If we should retain the concept of ideology, it is rather because it focuses our attention on the bidirectionality of practice, the materiality and externality of belief: not only on the use of the cultural tools by the Deweyan believers, but the use of the Deweyan believers by the tools. From this critical vantage and in view of Foucault's immense contribution to it, Rorty's "end of ideology" position is revealed to be thoroughgoingly ideological in a somewhat different and quite specific sense: its strategic misrecognition of ideology as an affair of thought only, a misrecognition that affirms the freedom of individual self-fashioning and ironist redescription (including, of course, Žižek's "lurid fantasies") while leaving wholly undisturbed the external and material "reality" of this liberal fantasy.

One such reality is the network of legal, civil, and medical institutions in liberal society that Rorty declares in fine working order in his notorious response to Clifford Geertz's critique of Rortyan pragmatism. Rorty answers Geertz's example of ethnocentrism—in which an alcoholic Native American is allowed to continue treatment on a kidney machine even though he refuses his doctors' advice to stop drinking, and so dies a few years later—by maintaining that it shows "our liberal institutions functioning well and smoothly" (*ORT* 204): "The whole apparatus of the liberal democratic state...insured that once the Indian had the sense to get into the queue early, he was going to have more years in which to drink than he would otherwise have had" (204). For Geertz, the example illustrates that "nobody in this episode learned very much about either themselves or about anyone else," and that "the whole thing took place in the dark," because what the doctors in question lacked was "knowledge of the degree to which he [the Indian] has earned his views" and "comprehension of the terrible road over which he has had to travel to arrive at them and of what it is—ethnocentrism and the crimes it legitimates—that has made it so terrible" (quoted in *ORT* 205). Rorty responds: "the fact that lots of doctors, lawyers, and teachers are unable to imagine themselves in the shoes of their patients, clients and students does not show that anything is taking place in the dark. There is light enough for them to get their job done, and to do it right" (205).

What seems clear here is that Rorty's response to Geertz's example is *wholly* contained within a technocratic functionalism—within what Jean-François Lyotard calls a "positivist pragmatism, which, beneath its liberal exterior, is no less hegemonic than dogmatism."²⁵ Like Ronald Reagan's homeless, who are "free" to sleep on the street if they so wish, Rorty's Indian is free to drink himself to death and thereby reconfirm a pernicious marginality whose conditions of social production are left bracketed, as if there were no relation at all between liberal ethnocentrism and the history of Native Americans' oppression and the social pathologies it has generated, no systematic connection at all between how well the legal system functions and for whom, depending on the citizen's economic standing or race. But what a Foucauldian analysis—a more Foucauldian *pragmatism*—helps us see is that the liberal technocratic functionalism at work in Rortyan pragmatism is, despite its patina of liberal concern *and* its philosophical antifoundationalism, a perfect instance of what Foucault calls "panoptical reason," "one that is self-contained and nontheoretical, geared to efficiency and productivity," which seems "to pose no standard of judgment or to follow any particular program."²⁶ Rorty may disavow in the realm of thought what humanist Reason *is*, but Foucault's more materialist pragmatism draws our attention to the fact that Rortyan pragmatism leaves intact what humanist Reason *does*.

> Rorty tells us that his preference for Dewey over Foucault is that

> > in Dewey's hands, the will to truth is not the urge to dominate but the urge to create, to "attain working harmony among diverse desires." This may sound too pat, too good to be true. I suggest that the reason we find it so is that we are convinced that liberalism requires the notion of a common human nature, or a common set of moral principles which binds us all, or some other descendent of the Christian notion of the Brotherhood of Man. So we have come to see liberal social hope—such as Dewey's—as inherently self-deceptive and philosophically naive. We think that, once we have freed ourselves from the various illusions which Nietzsche diagnosed, we *must* find ourselves all alone, without the sense of community which liberalism requires. (*Consequences* 207)

The very point of Foucault's emphasis on power and the idea of Man, however, is to focus our attention on the fact that in projects such as Rorty's, "liberal social hope" *is* "self-deceptive and philosophically naive," not because of its supposed foundationalism, but rather because it glosses over the contradiction we encounter time and again in Rorty's own work between pluralist, antirepresentationlist recognition of difference in the epistemological realm, and the status quo competitive individualism—replete with all of the judicial and economic apparatuses that secure its privileges—that *undermines* "community" in the material, political realm. For Foucault, to critique liberal foundationalism (as Rorty does with Habermas) while leaving the institutions and structures of liberal society unchanged is merely to show how little philosophy of a certain brand matters.

The political point for Foucault, we might say, is not so much that after the death of man we "find ourselves alone," but rather that some forms of "aloneness" are better than others—a fact that Rorty's undifferentiated "we" merely skirts. It is for this reason that Foucault reminds us that it is necessary

> to determine what "posing a problem" to politics really means. R. Rorty points out that in these analyses I do not appeal to any "we"—to any of those "we's" whose consensus, whose values, whose traditions constitute the framework for a thought and define the conditions in which it can be validated. But the problem is, precisely, to decide if it is actually suitable to place oneself within a "we" in order to assert the principles one recognizes and the values one accepts; or if it is not, rather, necessary to make the future formation of a "we" possible, by elaborating the question. Because it seems to me that the "we" must not be previous to the question; it can only be the result—and the necessarily temporary result—of the question as it is posed in the new terms in which one formulates it.[27]

For instance, Foucault's reading of the drunken Indian example would instead underscore (as in his work on marginality in general) how this is a classic instance of the *productive* power deployed by liberal society's "technologies of normalization," which not only isolate anomalies in the social body but also normalize them by "helping" and "managing" them with "purportedly impartial techniques" of "corrective or therapeutic procedures" (Rabinow, "Introduction" 21).

This does not mean that Foucault would presume to speak *for* the Indian—indeed, he spent his entire career resisting that view of the relationship between intellectuals and power, nowhere more so than in his famous dialogue with Deleuze in *Language, Counter-Memory, Practice*.[28] Instead, his aim in his work on marginality is to draw attention to

> unqualified, even directly disqualified knowledges (such as that of the psychiatric patient, of the ill person ... of the delinquent, etc.) ... a differential knowledge incapable of unanimity and which owes its force only to the harshness with which it is opposed by everything surrounding it—[it] is through the reappearance of this knowledge, of these local popular knowledges, these disqualified knowledges, that criticism performs its work.[29]

For Foucault, then, the intellectual does not have a *representational* relationship to the people in either the epistemological or the political sense; he is neither one who accurately reflects and mirrors the truth of a constituency back to it in more refined form, nor one who represents (as in "representational" democracy) the interests of a constituency in the culture at large. For Foucault, this "universal" intellectual has given way to what Foucault calls the "specific" intellectual. "It is in this context," Smart writes,

> that Foucault has suggested that the modern intellectual has a "three-fold specificity," namely of class position, of conditions of life and work associated with intellectual activity, and of "the politics of truth in our societies"; and it is in relation to the latter, that is, struggles and conflicts around the question of truth, that a politicization of intellectuals has become most apparent. In short, Foucault's position is that intellectuals are inextricably involved in a struggle over "the status of truth and the economic and political role it plays" and that the option before radical intellectuals is not that of "emancipating truth from every system of power (which would be a chimera, for truth is already power) but of detaching the power of truth from the forms of hegemony, social, economic, and cultural, within which it operates at the present time." ("Politics of Truth" 165–66)

This helps to elucidate *both* Foucault's differences with Rorty regarding the work of the intellectual *and* why those differences do not lead Foucault to "speak for" those marginalized by the liberal society whose "power of truth" Rorty legitimizes even as he disavows its "truth" philosophically. The aim of Foucault's genealogical critiques, Smart writes, is "to identify strengths and weaknesses in the networks of power, to provide in short, tools or 'instruments for analysis' and to leave the question of tactics, strategies, and goals to those directly involved in struggle and resistance" (167).

But to let matters rest here would be to give a purely negative account of Foucault's view of the politics of theory, and to ignore a certain undecidability in Foucault's approach to examples such as Geertz's Indian. After all, one could imagine a perfectly plausible Foucauldian reading of that instance which would see the Indian's refusal to give up his alcoholism not as a sign of repression and normalization but rather as a kind of "microresistance" to the system's management strategies; for, as Foucault writes in his analysis of "biopower" in the first volume of the *History of Sexuality*:

> It is over life, throughout its unfolding, that power establishes its domination; death is power's limit, the moment that escapes it; death becomes the most secret aspect of existence, the most "private." It is not surprising that

> suicide...became, in the course of the nineteenth century, one of the first
> conducts to enter the sphere of sociological analysis; it testified to the indi-
> vidual and private right to die, at the borders and in the interstices of power
> that was exercised over life. (*Foucault Reader* 261)

By this definition, the Indian's commitment to his alcoholism might well be seen as a sign of power's limit.[30] And the political undecidability here is not simply a matter of Foucault's well-known reluctance to be specific about what forms resistance might take. Rather, it is an undecidability that Foucault seems dedicated to making a permanent feature of his thinking, and as such it is an undecidability that points toward a *positive* characterization of what Foucault has called the "ethics of thought"—a force or process of thinking that supersedes any specific political problematic. As he puts it in "The Subject and Power":

> Maybe the target nowadays is not to discover what we are, but to refuse what we are. We have to imagine and to build up what we could be to get rid of a political "double bind," which is the simultaneous individualization and totalization of modern power structures. The conclusion would be... to liberate us both from the state and from the type of individualization which is linked to the state. We have to promote new forms of subjectivity through refusal of this kind of individuality which has been imposed on us for several centuries. (*Foucault Reader* 22)

As James Bernauer has argued, this desire to move away from the self that one is, this commitment to thought as a process of exploring what Foucault calls "new forms of subjectivity" and "limit experiences" (*Remarks on Marx* 27, 31), is an abiding concern for Foucault both early and late. This is most pronounced in Foucault's interviews, where he asks with some regularity, "What can be the ethic of an intellectual if not that: to render oneself permanently capable of getting free of oneself?" (quoted in Bernauer, *Michel Foucault's Force of Flight* 179). And this desire for thought as "freedom in relation to what one does, the motion by which one detaches oneself from it" (*Foucault Reader* 388), occasionally appears explicitly in Foucault's written work as well, as at the end of his introduction to *The Archaeology of Knowledge*, where he envisions his writing as "a labyrinth into which I can venture, in which I can move my discourse, opening up underground passages...in which I can lose myself and appear at last to eyes that I will never have to meet again. I am no doubt not the only one who writes in order to have no face. Do not ask who I am and do not ask me to remain the same."[31] Beneath Foucault's genealogical engagements with marginality, materiality, and disciplines one finds time and again the invocation of an ethics of philosophy reiterated at the end of Foucault's career, in the preface to volume 2 of the *History of Sexuality:* "it would probably not be worth the trouble of making books," Foucault writes, "if they failed to teach the author something he hadn't known before, if they didn't lead to unforeseen places, and if they didn't disperse one toward a strange and new relation with himself. The pain and pleasure of the book is to be an experience" (*Foucault Reader* 339).

Statements such as these appear too consistently—and are too strategically placed—in Foucault's work to be dismissed as mere "rhetoric" or "posturing." Indeed, as both Bernauer and John Rajchman have argued, it is crucial to any understanding of Foucault that we recognize the importance of

> thought's fundamental experience of itself as a force of flight. . . . [T]he essential characteristic of his thought is precisely this dynamic movement of relentless questioning that refuses to remain within one specific area of study and draw out fully the implications of a particular investigation. Foucault's style mirrors the fundamental urgency of his thought, which is less to convince than to agitate, to compel a desire for flight, to afflict the reader with a pressure or force. (Bernauer, *Michel Foucault's Force of Flight* 6)

It is at this juncture, of course, that Foucault's work is most open to the charges of "anarchism" and "self-absorption" often leveled at him by his liberal and leftist critics.[32] But, as Rajchman argues, those charges take for granted a view of philosophy explicitly under critique in Foucault's body of work. Foucault's critique does not aim to disclose rational norms for a totalizing analysis of society; instead, it is a "constant 'civil disobedience' within our constituted experience," in which the "questioning of anthropologism turns into an ethic of free thought: in suspending universalist narrative and anthropological assurance about an abstract freedom," Rajchman continues, "Foucault directs our attention to the very concrete freedom of writing, thinking and living in a permanent questioning of those systems of thought and problematic forms of experience in which we find ourselves."[33] It is this Foucault who is indeed reminiscent of American pragmatism—not of its liberal complacency, but its subterranean wildness and negativity found in the "anarchistic" side of William James and the "whimsical" circumambulations of Emerson.[34] This is the sense of pragmatism, we will remember, foregrounded by Stanley Cavell's sense of philosophy (and of Emerson) as "finding" rather than "founding," as a task of "transience" and "onwardness," a flight away from "thinking as clutching," a process of moving on that is crucial to the project of "moral perfectionism" and, in his view, to democracy. It is this view of philosophy, and this sort of Foucault, that is brought to the fore in the work of our next subject, Gilles Deleuze.

A Pragmatics of the Multiple: Foucault with Deleuze

What most conspicuously links the Rorty of *Philosophy and the Mirror of Nature* to the Foucault of *Discipline and Punish* is a thoroughgoing critique of the cultural work done by optical metaphors and cartographies. But it would be a mistake to collapse the figure of vision *tout court* into Panopticism, to see the visual as always already a form, as it were, of the Look. Or that is the position, at least, of Gilles Deleuze in his difficult and important book on Foucault, which argues that the signal advance of *Discipline and Punish* lies not in its Orwellian vision of the "hard" totality of Panoptical society, but rather in its potentially liberatory mapping of "new coordinates for praxis."[35] In this, Deleuze's Foucault emerges as a pragmatist par excellence, who with Deleuze (and his collaborator Félix Guattari) may be distinguished from the postmodernism of both Baudrillard and Lyotard by means of the statement that opens Deleuze's preface to the English edition of his *Dialogues:* "I have always felt than I am an empiricist, that is, a pluralist"—an equivalence between terms that Deleuze derives from Whitehead's redefinition of empiricism: that "the abstract does not explain, but must itself be explained"; that "the aim is not to rediscover the eternal in the universal, but to find the conditions under which something new is produced (*creativeness*)"; that the aim of philosophy is to analyze "the states of things"—which "are neither unities nor totalities, but *multiplicities*"—"in such a way that non-pre-existent concepts can be extracted from them."[36]

Deleuze's explicitly pragmatist and pluralist philosophical undertaking situates itself against what he calls "state philosophy," where

> thought borrows its properly philosophical image from the state as beautiful, substantial or subjective interiority. It invents a properly spiritual State, as an absolute state, which is by no means a dream, since it operates effectively in the mind. Hence the importance of notions such as universality, method, question and answer, judgment, or recognition, of just correct, always having correct ideas. Hence the importance of themes like those of a republic of spirits, an enquiry of the understanding, a court of reason, a pure "right" of thought, with ministers of the Interior and bureaucrats of pure thought. (*Dialogues* 13)[37]

Deleuze's pluralism is thus pragmatist in a very specific sense: not "'passive pragmatist' measuring things against practice"—or against what Lyotard calls the "performativity principle" of a "positivist" pragmatism, where utility and results are all that count (*Postmodern Explained* 66)—but " 'constructive' pragmatist whose aim is 'the manufacture of materials to harness forces, to think the unthinkable.' "[38]

As we shall see, it is precisely here, in its giddy invocation of "the unthinkable," that Deleuze's vision of pluralism reveals itself to be dependent on what we might call a romance or fantasia of "the Outside," that "turbulent, stormy zone" of a raging "battle," a "teeming mass" of "savage particular features" "where one can live and in fact where Life exists *par excellence*."[39] As Fredric Jameson has shrewdly observed, this desire to "shed our defenses and give ourselves over absolutely to this terrifying rush of the non-identical is of course one of the great ethical fantasy-images of the postmodern,"[40] and it is a fantasy-image whose problems for many critics on the left are at least threefold. First (to borrow Stephen Best and Douglas Kellner's formulation), it "uncritically assimilate[s] the modernist ethos of incessant self-transformation, becoming, and psychic instability," and thus participates in a "decentering of ethics in favour of aesthetics that is typical of postmodern theory."[41] Second, as Jameson and others have pointed out (though this is less true of Deleuze's book on Foucault than of the collaborations with Guattari), Deleuzian pluralism short-sells the crucial political issues of intersubjectivity and the potentially empowering forms of identity and stability.[42] And third, "it is not clear," as Best and Kellner put it, "that this position radically breaks from capitalist and consumerist behavior" (*Postmodern Theory* 107). In this regard, it would be overly simple, but not altogether wrong, to agree with Jameson that "the theorists of French poststructuralism simply change the valences on the old descriptions of Adorno, Horkheimer, and Marcuse, so that what used to be denounced as commodification is now offered by Deleuze and Guattari as the consciousness of the ideal schizophrenic, the 'true hero of desire.' "[43] And it is hard to see how that hero—whether a reincarnation of the delirious consumer of capitalist culture or not—could engage in collective political practice with others.

At the same time, however, what needs to be understood is that this very notion of practice itself (as the Foucauldian "ethics of thought" suggests) is what is centrally under dispute in Deleuze and Foucault (as is, for that matter, the vision of "theory" and philosophy that usually accompanies it)—and not least of all because of the skepticism toward it generated by the well-known events of May 1968 in France, events that are everywhere between the lines in the important Deleuze/Foucault dialogue, "Intellectuals and Power." As Best and Kellner point out in their excellent overview of postmodern theory, for Deleuze, Guattari, and Foucault, the events of May 1968 provided an important lesson in the limits and failures of these traditional "macroperspectives" of theory and practice, which could

only see the social upheaval in France at that moment as "diversionary" or "imma-
ture," because they failed to understand that the truly revolutionary political potential of the moment lay beyond the strict purview of class contradictions and crises
of capital (100–101). In this light, what is crucial about the Deleuzian intervention—and this is especially evident in, but by no means limited to, his work with
Guattari—is its recognition of the crucial *micropolitical dimension* of capitalist culture. Deleuze understands (as does Guattari)

> that genuinely radical politics cannot simply make rational appeals to sub-
> jects concerning the nature of their oppression and provide cogent reasons
> why they should overthrow their oppressors....A traditional rationalistic
> macropolitics leaves the terrain of desire, culture, and everyday life uncon-
> tested, precisely the spaces where subjects are produced and controlled, and
> where fascist movements originate. (Best and Kellner, *Postmodern Theory* 94)

To recognize this fact is to at least partly dispense with the charge of complicity
with commodification and consumerism often leveled at Deleuze and Guattari, because with regard to the imbrication of desire and investment with capitalism,
"There is no getting outside it," as Brian Massumi rightly argues. If this is so, then
the path toward progressive political and social change, at least from a Deleuzian
point of view, cannot consist of "lamenting the loss of such 'traditional values' as
belief and sincerity and reverting to moralism, or mourning the 'death of the subject,'" but rather "in taking the inventive potential released by capitalism so far that
we become so other as to no longer act in the perceived 'private' interests of a separate Self that we have in any case ceased to be." This is a very different way, in
other words—a way foreign to traditional notions of practice—to "embrace our
collectivity" (Massumi, *User's Guide* 140–41). In view of the Deleuzian micropolitical
perspective, collectivity happens not by forging traditionally "grounded" alliances
with other subjects, by joining a bloc whose coherence depends on paring away
anything *other* than the singular "identity" (of class, of gender) that binds the group
together, but rather by recognizing—as the work of Donna Haraway and Bruno
Latour so powerfully does—that the imaginary "Self" of capitalist/patriarchal culture is *already* a concrete "superindividual composed of a multitude of subindividu-
als" (Massumi, *User's Guide* 81). Capitalism might take advantage of this multisub-
ject, of course, but it is also true that such a subject "always has the potential to
reconnect with its impersonality to become a subject-group.... Since a person is
only as stable as its constituent contractions—that is, metastable—it can be pre-
cipitated into a crisis state despite its best intentions" (81).

This rejection of philosophy as the rational consensus that makes collective practice possible is surely behind Deleuze and Guattari's assertion in their final collaboration, *What Is Philosophy?* (and here we cannot fail to detect, I think, jabs at both Habermas and Rorty), that "it does no credit to philosophy for it to present itself as a new Athens by falling back on Universals of communication that would provide rules for an imaginary mastery of the markets and the media (inter-subjective idealism).... The first principle of philosophy is that Universals explain nothing but must themselves be explained."[44] As with Foucault's "ethics of thought," the object of Deleuzian philosophy is above all "to create concepts that are always new" (5),[45] but "The idea of a Western democratic conversation between friends has never produced a single concept" (6).

And when one asks, as Deleuze and Guattari do in their last work, "Why, through what necessity, and for what use must concepts, always new concepts, be created? and in order to do what?" the answer is a thoroughgoingly pragmatic one—but pragmatic in a Foucauldian rather than a Rortyan sense. As Michael Hardt argues in his study of Deleuze, the essentially pragmatic character of Deleuze's thought makes it rather fruitless "to attempt a general definition of the politics of poststructuralism, or even of the politics of Deleuze's philosophy. It is more appropriate and more productive to ask ourselves, What can Deleuze's thought afford us? What can we make of Deleuze? In other words, what are the useful tools we find in his philosophy for furthering our own political endeavors?"[46]

The "concept" for Deleuze is thus precisely the opposite of what it was for Adorno.[47] As Todd May summarizes it, it "is not a representation in any classical sense. Rather, it is a point in a field—or, to use Deleuze's term, on a 'plane'—that is at once logical, political, and aesthetic. It is evaluated not by the degree of its truth or the accuracy of its reference, but by the effects it creates within and outside of the plane on which it finds itself."[48] Deleuze's unabashed metaphysical bent may then be seen as a kind of ad hoc supporting structure or scaffolding enabling the construction of these planes, which in turn serve a fundamentally pragmatist relation to philosophy.[49] As Hardt puts it, in Deleuze, "Ontological speculation prepares the terrain for a constitutive practice; or rather, after ontological speculation (as *Forschung*) has brought to light the distinctions of the terrain, this same terrain is traversed a second time in a different direction, with a different bearing, with a practical attitude (as *Darstellung*)." Thus, "we can give a Deleuzian reading to Lenin's insight. 'Without theory, no revolutionary practice': Without theory there is no terrain on which practice can arise, just as inversely, without practice, there is no terrain for theory. Each provides the conditions for the existence and development of the other."[50]

All of which helps to clarify a few important points we need to understand about reading Deleuze (and Deleuze's Foucault). For Deleuze, philosophy is above all *experimental thinking with a pragmatic rationale*. It makes very little sense — indeed, it is essentially a waste of time — to read Deleuze with expectations for thinking and its productivity that one would bring to a Kantian or Cartesian or even (especially) Hegelian philosophy. (Indeed, the thrust of much of Deleuze's work has been to locate an alternative tradition to this type of philosophy in the work of Bergson, Leibniz, Spinoza, and other "outsider" figures.) To gain anything from reading Deleuze (this is by no means limited to Deleuze, of course), you must be willing to play his game, to go along, in a kind of philosophical negative capability, with his redefinition of philosophical thinking and his redirection of familiar problems and concepts. And this is true as well of what we could characterize as the political dimension or "relevance" of Deleuze's thought, which will strike many readers of Deleuze's work after *Anti-Oedipus* as oblique at best. Here again, what is important to keep in mind is that Deleuze's pragmatism resides in no small part in its refusal to see its vocation as providing "grounds" or "frames" for any particular, prespecified political program or form of practice. Those who expect the political relevance of philosophy to consist in its ability to give the causes of political problems and then specify their solutions will find Deleuze an a- or even antipolitical thinker.

But we should keep in mind, I think, that Deleuze's philosophical and political intervention takes place at a different (I am tempted to say, in completely un-Deleuzian fashion, "deeper") level than that. Deleuze's thinking is concerned instead with the *conditions of possibility* (he would say, I think, the conditions of *existence*) for politics — that is, with the conditions and dynamics under which specific forms of power and domination persist, and the forms of resistance it is possible to imagine that they generate, in the ceaseless struggle between exclusionary, identitarian social forms (be they of economics, gender, sexuality, or whatever) and their own outsides. In this deeper sense, as Hardt points out, Deleuze "can help us develop a dynamic conception of democratic society as open, horizontal, and collective." "Deleuzian being," he continues,

> is open to the intervention of political creations and social becomings.... The power of society, to translate in Spinozian terms, corresponds to its power to be affected. The priority of the right or the good does not enter into this conception of openness. What is open, and what links the ontolog-

ical to the political, is the expression of power: the free conflict and composition of the field of social forces. (120)

In political terms, then, Deleuze's thought will never tell us what to do, or even
what always already counts as "genuinely" political and what does not; it is rather a
pragmatic set of insights, interventions, and tools for prying open the space of social forces and theorizing them, the better to explore forms of social difference and
resistance.

      This will help to clarify why, for Deleuze, Foucault's pragmatism
lies not only in his break with the Marxist theory of the relationship between politics and power—and especially with Marxism's "complicity about the state" (*Foucault* 30) as a privileged apparatus of power—but also, and more important, in its
theorization of the irreducible difference between "the visible" (or nondiscursive)
and "the articulable" (or discursive) (31). To misunderstand Foucault's critique of
Panopticism as a reading of the visual as such would be to fall prey to what Deleuze
calls "Foucault's great fiction" (and a very common reading of him at that): that
"the world is made up of superimposed surfaces, archives or strata. The world is
thus knowledge" (120)—that is, it is utterly panoptical. It is also to miss the central
linkage between contingency, pragmatism, and resistance in Deleuze's Foucault: that
the repetition and reproduction of statements, archives, and all that comes from the
strata of historical "knowledge" takes place in a context of multiplicity and contingency, which is simply to say that pure repetition is impossible, that repetition always takes place with a difference—and from that fact springs what Deleuze calls
"microagitations," the emergence of new forms of thought and practice. The crucial pragmatic point of Foucault's work after the *Archaeology* is that "the final word
on power is that *resistance comes first*." While relations of power operate within the
forms of "knowledge," within the "diagrams" and "abstract machines" that link the
visible and the discursive in a circuit that power traverses, relations of *force* operate
beneath and, as it were, before those relations—hence "resistances necessarily operate in a direct relation with the outside from which the diagrams emerge" (89).
"This," Deleuze concludes, "is the whole of Foucault's philosophy, which is a pragmatics of the multiple" (83–84).

      Here, Deleuze offers a revisionist reading of Foucault's analytic
of power that meets Foucault halfway, as it were. As Best and Kellner suggest, the
earlier Deleuze had, with Guattari, critiqued Foucault's account of power by arguing
that "Power is epiphenomenal to the flow of desire. Second, and consequently, the
lines of flight are fundamentally positive and creative, rather than lines of resistance

or counter-attack." Thus, desire is seen in the early Deleuze and Guattari collaboration as "purely affirmative, and not a desire to resist another force," and "the philosophy of authentic multiplicities" depends on multiplicities being "analyzed without being related to a lost unity or totality" (*What Is Philosophy?* 101). In *Foucault*, Deleuze leaves aside the language of desire and the psychoanalytic problematic, thus moving toward Foucault's analytics, while at the same time suggesting that Foucault's new understanding of power does justice to the multiplicity of forces at work in the social field.

Here, according to Deleuze, *Discipline and Punish* marks a crucial advance beyond *The Archaeology of Knowledge*, which assigned only a negative, epiphenomenal role to the nondiscursive "environments" ("institutions, political events, economic practices and processes") in which statements operate (*Foucault* 31). "It is here," Deleuze writes,

> that *Discipline and Punish* poses the two problems that *The Archaeology* could not raise because it remained tied to Knowledge, and the primacy of the statement in knowledge. On the one hand, outside forms, is there in general a common immanent cause that exists within the social field? On the other, how do the assemblages, adjustments and interpenetration of the two forms come about in a variable way in each particular case? (33)

But even as *Discipline and Punish* moves beyond what might be called the discursive formalism of the *Archaeology* and its tie to "Knowledge" (in contrast to that dynamic and productive process Deleuze will call "Thought"), it immediately raises another question: If the visible and the articulable are irreducibly different assemblages of heterogeneous, multiple elements, then how is it that their operations are so often coordinated with devastating pragmatic consequences in the social field? After all, there is certainly a very tight coordination between the set of statements that constitute the penal code and the set of visibilities put to service in the Panopticon. So even though the visible and the articulable are irreducible in their difference, how do we explain their "coadaptation"?

Two ways *not* to explain it, according to Deleuze, are by reference to the theory of ideology and by recourse to a semiotic or signifier-based model of meaning. As for the first of these, Foucault's pragmatism rests on his "new functionalism" (25) that "throws up a new typology which no longer locates the origin of power in a privileged place, and can no longer accept a limited localization" (26) because, as *Discipline and Punish* demonstrates, "discipline cannot be identified with any one institution or apparatus precisely because it is a type of power, a technology, that traverses every kind of apparatus or institution, linking them, prolonging them, and making them converge and function in a new way" (26). And this means in turn that "Power does not come about through ideology" (28). "Power"—as Deleuze would remind Rorty—"'produces reality' before it represses. Equally it produces truth before it ideologizes, abstracts or masks" (29). Like desire in the early Deleuze and Guattari, power "is fundamentally positive and productive in nature," operating "out of the productive plenitude of its own energy which propels it to seek ever new connections and instantiations"; it is, to use the language of *Anti-Oedipus*, a "dynamic machine" (Best and Kellner, *Postmodern Theory* 86). What Foucault's new pragmatism makes clear is that "repression and ideology explain nothing but always assume an organization or 'system' within which they operate, but not vice versa. Foucault does not ignore repression and ideology; but as Nietzsche had already seen, they do not constitute the struggle between forces but are only the dust thrown up by such a contest" (Deleuze, *Foucault* 29).[51]

We have already seen that Foucault explicitly rejects the theory of ideology for three reasons: it presumes something on the order of a constituent subject; it indulges the "repressive hypothesis"; and it is inescapably linked to a reflectionist, base-superstructure model. But what Deleuze's reading helps us understand is something like the ur-motive anchoring Foucault's three-pronged critique. As Massumi characterizes it, "Power can be conceived as language-driven but not language-based. Its functioning cannot fully be explained by recourse to a concept of ideology as formative agent of speech and belief....An ideological statement is more a precipitate than a precipitator" (*User's Guide* 154 n. 45)—it is, indeed, the "dust" thrown up by "the struggle between forces" that are not themselves language-based.

This leads us, then, to the second way *not* to answer the question of the "coadaptation" of the discursive and the nondiscursive, and that is by recourse to semiotic or signifier-based models. For, as Deleuze and Guattari point out in *A Thousand Plateaus*, "Signifier enthusiasts take an oversimplified situation as their implicit model: word and thing. From the word they extract the signifier, and from the thing a signified in conformity with the word, and therefore subjugated to the signifier. They operate in a sphere interior to and homogeneous with language."[52] On the other hand, neither are expressions "reducible to base-superstructure. One can no more posit a primacy of content as the determining factor than a primacy of expression as a signifying system" (68). Instead, we must recognize that all expressions are always involved in what Deleuze and Guattari call a "double articulation," since both "form of expression" and "form of content" are themselves doubly embedded in relatively internal relations of form and relatively external relations of substance. As Deleuze and Guattari put it, each "has a code *and* a territoriality; therefore each possesses both form and substance" (41). To borrow Massumi's example, we can see that in the case of woodworking, "Expression has no more monopoly on form than content does on substance. There is substance on both sides: wood; woodworking body and tools. And there is form on both sides: both raw material and object produced have determinate forms, as do the body and the tools. The encounter is between two substance/form complexes," and not just between the *simple* elements in the substance/form or content/expression dualisms (*User's Guide* 12). As Deleuze and Guattari explain the relationship, "There is never correspondence or conformity between content and expression"—as in the Saussurean scheme, in which the signifier unites a "sound-image" and a "concept"—"only isomorphism with reciprocal presupposition" (*Thousand Plateaus* 44). They "are rather, as it were, two non-parallel formalizations, the formalizations of expression and the formalizations of content, such that one never does what one says, one never says what one does, although one is not lying, one is not deceiving or being deceived" (*Dialogues* 71).

It is this double complexity of "reciprocal presupposition," this "double articulation," that is overlooked by signifier-based models of meaning, for, as Massumi points out, "By bracketing the statement's real conditions of social emergence," signifier-based models

> cut it off from its efficient cause: the overall abstract machine that pragmatically determines the substance as well as the form of both content and expression in their double articulation. Theories of the signifier reduce language to expression and expression to its form. In so doing, they unmoor language from its "vertical content" [its relation to the nondiscursive or the visible], from the realm of virtuality constituting its real becoming as a hand-to-hand combat of energies. (*User's Guide* 44)

As Massumi argues, the reconnection of the discursive to the nondiscursive is what both Baudrillardian and Lacanian approaches neglect to do, because they "reduce 'vertical content' to a signified" (45)—rather than recognizing that the form of content has its own vertical and horizontal relations of both "code" and "territoriality"—and then they untether *that* from the plane of "horizontal" relations of force, of nonintentional interactions and aleatory events, by declaring the horizontal plane "a 'referent' lying irretrievably outside language understood as a closed system or two-dimensional form of interiority" (45).[53]

Instead, Deleuze argues, we should follow Foucault's pragmatic method, which shows that the prison as a form of content related to other forms of content (schools, hospitals, factories) "does not refer back to the word 'prison' but to entirely different words and concepts, such as 'delinquent' and 'delinquency,'" which are themselves the "form of expression in reciprocal presupposition with the form of content 'prison'" (*Thousand Plateaus* 66). To put it another way, the prison as a social form (of content) does not refer for its meaning — much less for what we might call its "reality" — to the signifier (or form of expression) "prison," but rather to other forms of expression (deliquency, vagrancy, deviance, etc.), whose forms of content are *other* specific relations to forces, energies, and spaces (circulation and reiteration in vagrancy's relation to spaces, for example, versus the analytical plotting and distribution of it in the prison). For Deleuze, Foucault's pragmatism therefore resides in part in his recognition of the inadequacy of the signifier-based — or, as in *The Archaeology of Knowledge*, statement-based — strategy, in his insistence on the irreducible difference of the articulable and the visible, the form of expression *and* the form of content, and, within each of those domains, on the vertical and horizontal relations of meaning.

To return to our question, then: how are these different domains coordinated and coadapted? Foucault's answer is that it is the "diagram" — whose privileged instance in *Discipline and Punish* is Panopticism — that fulfills this function. It is the diagram that sets up relations of correspondence between specific points in the form of expression and the form of content, and thereby imposes "a particular conduct on a particular human multiplicity" (34). It is, Deleuze writes,

> no longer an auditory or visual archive but a map, a cartography that is co-extensive with the whole social field. It is an abstract machine. It is defined by its informal functions and matter and in terms of form *makes no distinction between content and expression, a discursive formation and a non-discursive formation.* It is a machine that is almost blind and mute, even though it makes others see and speak. (*Foucault* 34; my emphasis)

The diagram — *because* it is "informal," because it pays no attention to formal specificity, to the distinction between discursive and nondiscursive, the form of expression and the form of content, and thus assumes complete isomorphism between the statement and its nondiscursive enactment — constitutes the power network of a society and coordinates the coadaptation of these different domains.

But the specificity of Foucault's pragmatism — his posing of "new coordinates for praxis" — lies not only in his articulation of the irreducibility of the discursive and the visible, but also, and more broadly, in his theorization of the fact that Power and Knowledge (as opposed to what Deleuze calls "Forces" and "Thought") are founded on an abyss, structured across a void or what Foucault calls a "non-place" that lies between the discursive and the articulable. As Massumi explains it, "If meaning is the in-between of content and expression, it is nothing more (nor less) than the being of their 'nonrelation'" (*User's Guide* 16), because—to stay with the woodworking example—"the interrelation of relations between the wood and the tool bears no resemblance to that between concepts, which bears no relation to that between phonemes or letters" (17). The power of the diagram is precisely to set up and maintain correspondences between these planes of articulation: "All knowledge," Deleuze writes, "runs from a visible element to an articulable one, and vice versa" (*Foucault* 39). "If knowledge consists of linking the visible and the articulable," he continues, "power is its presupposed cause; but, conversely, power implies knowledge as the bifurcation or differentiation without which power would not become an act" (39). Even as knowledge is a "diagramming" function, in other words, it always already presupposes the bifurcation or differentiation of domains that is power's raison d'être; if the two domains were always already one—if what we say could automatically produce what we see, or vice versa—then power would have no reason for being. As it is, "what we see never resides in what we say," as Foucault puts it (quoted in *Foucault* 66), and thus the visible and the articulable, far from being regularized under the regime of the signifier and the semiotic model, must "grapple like fighters, force one another to do something or capture one another" (67), a process "which bears witness to the fact that the opponents do not belong to the same space or rely on the same form" (68). Anyone who tries to make a table simply by standing in front of a log and uttering the word "woodworking" will quickly get the point; and anyone who can set up a correspondence between those two spaces and forms in a dynamic and productive relation via knowledge will understand what power is.

What gives the diagram or abstract machine its power, then—the fact that it can set up relations between the discursive and the visible, form of expression and form of content—is also what makes it vulnerable: it is structured across a "nonplace," and the meaning it makes possible is the "relation of a nonrelation." The political promise of Foucault's philosophical pragmatism for Deleuze is to disclose both this power and this fragility, to provide an anatomy of the unifying and regularizing work of the diagram or abstract machine and then to show how it runs aground on the irreducible difference of the two poles that it spans. As Deleuze puts it:

> ultimately this realization and integration [of the diagram] is a differentia-
> tion: not because the cause being realized would be a sovereign Unit, but on
> the contrary because the diagrammatic multiplicity can be realized only and
> the differential of forces integrated only by taking diverging paths, splitting
> into dualisms, and following lines of differentiation without which every-
> thing would remain in the dispersion of an unrealized cause. (*Foucault* 37–38)

And thus, Deleuze concludes, "between the visible and the articulable a gap or disjunction opens up, but this disjunction of forms is the place"—or more properly
"nonplace"—"where the informal diagram is swallowed up and becomes embodied
instead in two different directions that are necessarily divergent and irreducible.
The concrete assemblages are therefore opened up by a crack that determines how
the abstract machine performs" (38). What Deleuze tells us here in so many words
is that the price that the diagram pays, as it were, that knowledge and power must
always pay for their exercise, is the price of *embodiment*, of *materiality* and *multiplicity*, the risk of being "swallowed up" in the "nonplace" between the visible and the
articulable. "Thus there is no diagram," Deleuze writes, "that does not also include,
besides the points which it connects up, certain relatively free or unbound points,
points of creativity, change, and resistance" (44).[54] In our woodworking example, it
is possible that your diagammatic linkage of visible and articulable via knowledge,
and your consequent exercise of power, your mastery of forces, will suddenly run
aground when it discovers, say, not cherry or mahogany under the blade of the
block plane but a previously unknown, very different, more stringy material, oak,
which might force or make possible in its "free" or "unbounded" nature new techniques, practices, and destabilizations of knowledge.

In a somewhat different and more explicitly political register,
Massumi characterizes the dynamics of the diagram this way: "Disciplinary institutions
do the dirty work of transcendence. Their function is to see that a body is channeled
into the constellations of affect and orbits of movement set out for it by its assigned
category. The category is a map of habit, a coded image enveloping life's path, a blueprint for how a body will be cut" (*User's Guide* 114). "Bodies that fall prey to transcendence," he suggests, "are reduced to what seems to persist across their alterations.
Their very corporeality is stripped from them, in favor of a supposed substrate—
soul, subjectivity, personality, identity—which in fact is no foundation at all, but an
end effect, *the infolding of a forcibly regularized outside*" (112; my emphasis).

This testifies, then, to an incipient, disruptive multiplicity always
pressing at the edges of any diagram, any form of power/knowledge. As Deleuze
writes in perhaps the most important passage in *Foucault*:

The above study presented us with a dualism peculiar to Foucault, existing on the level of knowledge, between the visible and the articulable. But we must note that in general a dualism has at least three meanings: it involves a real dualism marking an irreducible difference between two substances, as in Descartes, or between two faculties, as in Kant; or it involves a provisional stage that subsequently becomes a monism, as in Spinoza or Bergson; or else it involves a preliminary distribution operating at the heart of a pluralism. Foucault represents this last case. For if the visible and the articulable elements enter into a duel, it is to the extent that their respective forms, as forms of exteriority, dispersion or dissemination, make up two types of "multiplicity," neither of which can be reduced to a unity. Statements exist only in a discursive multiplicity, and visibilities in a non-discursive multiplicity. And these two multiplicities open up on to a third: a multiplicity of relations between forces, a multiplicity of diffusion which no longer splits into two and is free of any dualizable form. (83–84)

We need to remember here that for Deleuze, "forces" should not be confused with "power," for the former "operate in a different space to that of forms, the space of the Outside, where the relation is precisely a 'non-relation,' the place a 'non-place,' and history an emergence" (86–87)—an "emergence" not only in the sense discussed by Foucault in his well-known essay "Nietzsche, Genealogy, and History," but also in the sense used by the systems theory we have already examined.[55] "Seeing is thinking," Deleuze writes, "and speaking is thinking, but thinking occurs in the interstice, or the disjunction between seeing and speaking. This is Foucault's second point of contact with Blanchot: thinking belongs to the outside in so far as the latter, an 'abstract story,' is swallowed up by the interstice between seeing and speaking" (87).

When this happens, "when words and things are opened up by the environment without ever coinciding, there is a liberation of forces which come from the outside" (*Foucault* 87), forces that compose "a battle, a turbulent, stormy zone where particular points and the relations of forces between these points are tossed about" (120). Given Deleuze's stratospheric abstraction, it might be helpful to think of the space of the outside as composed of "points" or "singularities" that are free and unstructured, that have not yet been brought into a "plane" of consistency, and that thus generate maximum force because they exist in maximum potential connectivity. If we think this dynamic on the terrain of subjectification, then we can see that "the multitude of individuals that contract to produce the person is reduced to the one-two-(three) of self-other-(phallus)," while in reality, what lies outside of Oedipal subjectivity is not "a protometaphysical 'confusion'" ("regression" to a "pre-Oedipal" body, "denial" of the "fact" of castration, and so on), but rather "an effective superposition of an unaccustomed range of pragmatic potentials" (Massumi, *User's Guide* 84–85) — the multiple possibilities of *what a body can do* when reasserted, in its full "reality," as the "outside" of the Oedpial diagram. Such a reframing helps clarify the pragmatic political potential of Deleuze's analysis of the "diagram" and the "outside"; as Massumi argues, "The liberal nation-state's ability to find an integrative response to perceptions of the outside is stretched to the limit when confronted by sexual minorities," because "successful becoming-woman, becoming-lesbian or -gay, becoming sadomasochistic, or becoming-boy lover, directly challenges the universal form of . . . 'democracy': Oedipal personhood itself" — a point shrewdly recognized by the New Right, which, "for all its apparent archaism, has been far more attuned than the traditional Left to the actual lines of force in late capitalist society" (127).

In the more analytical and ontological key that dominates Deleuze's *Foucault*, the "outside" is perhaps best thought of, as Massumi points out, in terms of *virtuality*, in which each singularity or point composes a myriad of possible states, both in the internal relations of its own basic elements and in its external relations to other points. No *actual* state of any point can at any given time effectively exhaust or express all of these potential states, and so when the actual state is realized, "some potential states drop out of each global state's actuality, but they go on quietly resonating in another dimension, as pure abstract potential" (*User's Guide* 65). The actual and the virtual are thus "coresonating systems"; "a physical system paradoxically embodies multiple and normally mutually exclusive potentials, only one of which is selected."[56] Hence, Massumi writes, "The virtual as a whole is the future-past of all actuality, the pool of potential from which universal history draws its choices and to which it returns the states it renounces. The virtual is not undifferentiated. It is *hyperdifferentiated*. If it is the void, it is a hypervoid in continual ferment" (*User's Guide* 66).[57] What this means, then, is that what Deleuze calls "force" in *Foucault* "is immanent to matter and to events, to mind and to body and to every level of bifurcation composing them and which they compose. Thus it also cannot but be experienced, in effect — in the proliferations of levels of organization it ceaselessly gives rise to, generates and regenerates, at every suspended moment" (Massumi, "Autonomy of Affect" 94).

Understanding the outside as virtuality will help us to elucidate how the concept of the outside functions in a curious essay that is everywhere behind the scenes in Deleuze's deployment of the term: Foucault's "Maurice Blanchot: The Thought from Outside." There, Foucault writes that in light of the outside,

> language is then freed from all of the old myths by which our awareness of words, discourse, and literature has been shaped. For a long time it was thought that language had mastery over time, that it acted both as the future bond of the promise and as memory and narrative;...In fact, it is only a formless rumbling, a streaming; its power resides in its dissimulation. That is why it is one with the erosion of time.[58]

To submit discourse to the challenge of the outside, then, involves "a listening less to what is articulated in language than to the void circulating between its words, to the murmur that is forever taking it apart," to the "non-discourse of all language" steadily, erosively at work in "the invisible space in which it appears" (25–26). Language takes place in a context of nonlanguage, thought in a context of the unthought, and what we hear in this "non-discourse" is the "murmur," the "rumbling" and "streaming," of virtuality that overtakes language itself, subjecting it to the "erosion of time" but *also* mobilizing it as a force of dissimulation and Deleuzian differentiation that opens up new possibilities for praxis and resistance.

This rendering of the relation between force, multiplicity, and philosophical concepts in terms of virtuality extricates Deleuze, at least in part, from a formidable philosophical dilemma. As Massumi puts it, although the realm of force and the outside in Deleuze

> is transcendental in the sense that it is not directly accessible to experience, it is not transcendent, it is not exactly outside experience either. It is immanent to it—always in it but not of it....Deleuze's philosophy is the point at which transcendental philosophy flips over into radical immanentism, and empiricism into ethical experimentation. The Kantian imperative to understand the conditions of possible experience as if from outside and above transposes into an invitation to recapitulate, to repeat and complexify, ground level, the real conditions of emergence. ("Autonomy of Affect" 94)

But the problem here lies precisely in the extreme tension between the commitment to "real conditions" on the one hand, and, on the other, to an "ethical experimentation" that, as we have already seen, will take place pragmatically, by the production of new philosophical concepts. The issue here is *not* simply one of choosing between the virtual and the actual, because the two, as we have seen, are always already coimplicated.[59] The problem is rather that the "real" Deleuzian multiplicities or singularities are, as May puts it, invoked as "placeholders for what lies beneath all qualities," "the positive differences that subtend all unities. For Deleuze," he continues, "they exist—or better, subsist—beneath sense, language, concepts, bodies, consciousness, in short beneath all phenomena of experience. They are unexplained explainers" that "escape all accounting" ("Difference and Unity" 46). What May puts his finger on here is nothing less than the fundamental assumption of Deleuze the ontologist, the Deleuze who takes for granted the Bergsonian "positive emanation of being" through ceaseless differentiation. Of course, May observes, "such a strategic move is bound to fail":

> Only a philosophy that finds difference on the surface rather than in a source beneath or beyond it—even when that source eventually becomes the constitution of the surface—can articulate a role for difference that possesses both coherence and normative power. In allowing a place, often a constitutive place, for positive differences that are not themselves already differences of a surface, Deleuze allows his thought to lean exclusively on one half of the intertwining that is necessary in order to prevent his fragile project from collapsing. (47)

It may be, as Michael Hardt argues, that Deleuze's aim is to "only accept 'superficial' responses to the question, 'What makes being possible?' ... There is nothing veiled or negative about Deleuze's being; it is fully expressed in the world. Being, in this sense, is superficial, positive, and full. Deleuze refuses any 'intellectualist' account of being, any account that in any way subordinates being to thought, that poses thinking as the supreme form of being" (*Gilles Deleuze* xiii). If this is so, then what it means, according to Hardt, is paradoxically that *practice* is the "foundation of ontology" (xiii); Deleuze thus raises "the theory of practice to the level of ontology," and thereby reorients the theory of practice "toward the ontological rather than the epistemological realm" (xv). But it is not clear how *practice* can be constitutive of ontology, because, first, that would require us to assume the banal and trivializing view that there is finally nothing *but* practice and, second, it raises the obvious question of how practice can constitute that which by definition subtends it, which always already exists (as the "positive emanation of being") before and besides practice itself. Deleuze *wants* to hold, as Hardt puts it, that "being is expressed always and everywhere *in the same voice*" (113), that being is thus "superficial." But the problem with this view is that this superficiality relies on a concept of being that is indeed "deep" and "hidden"—that is to say, an ultimately noncontingent presupposition—insofar as Hardt is right that "the dignity of being is precisely its power, its internal production—that is, the efficient causal genealogy that rises from within, the positive difference that marks its singularity. Real being is singular and univocal; it is different in itself. From this efficient difference at the heart of being flows the real multiplicity of the world" (114). Hardt thinks this shows that "Deleuze's ontology requires a materialist perspective because any priority accorded to thought would weaken the internal structure of being" (114). But how can giving priority to thought weaken being if being is what (Hardt's) Deleuze says it is? How can that which is noncontingent be contingent?

Folded but Not Twisted: Deleuze and Systems Theory

To address these dilemmas raised by Deleuze's reorientation of the problematic of the outside toward ontological ground, we need to view it alongside the relation of concepts that obtains for the systems theory of Maturana, Varela, and Luhmann. As Katherine Hayles lucidly describes it:

> The originary moment for the creation of a system, according to Niklas Luhmann, comes when an observer makes a cut. Before the cut—before any cut—is made, only an undifferentiated complexity exists, impossible to comprehend in its noisy multifariousness.... The cut helps to tame the noise of the world by introducing a distinction, which can be understood in its elemental sense as a form, a boundary between inside and outside. What is inside is further divided and organized as other distinctions flow from this first distinction, exfoliating and expanding, distinction on distinction, until a full-fledged system is in place.[60]

Now, what is most interesting is that this seems to be the view of the relationship between concepts and differences (or "singularities") at work in Deleuze's final collaboration with Guattari, *What Is Philosophy?*—a fact that would seem to bear out Hardt's insistence that we always pay attention to the evolution of Deleuze's thought, not only (for Hardt) from ontology to ethics and then to politics, but also (for my purposes) from politics to what I have characterized as Deleuze's pragmatics (Hardt xx). In *What Is Philosophy?* Deleuze and Guattari tell us that there are no concepts "possessing every component, since this would be chaos pure and simple," and that therefore "we find the idea of the concept being a matter of articulation, of *cutting and cross-cutting*. The concept is a whole because it totalizes its components" (16; my emphasis). "What is distinctive about the concept," they continue, "is that it renders components inseparable *within itself*. Components, or what defines the *consistency* of the concept, its endoconsistency, are distinct, heterogeneous, and yet not separable" (19). The concept thus "has no *reference:* it is self-referential, it posits itself and its object at the same time as it is created" (22).

This view of the concept is very close indeed to Luhmann's account of how an autopoietic system carries out *observations*. To briefly remind ourselves, Luhmann's position is that all observations are constructed atop a constitutive distinction that is paradoxical, and to which the observing system must remain "blind" if it is to engage in that observation *at all*. Hence:

> The source of a distinction's guaranteeing of reality lies in its own operative unity. It is, however, precisely as this unity that the distinction cannot be observed—except by means of another distinction which then assumes the function of a guarantor of reality. Another way of expressing this is to say the operation emerges simultaneously with the world which as a result remains cognitively unapproachable to the operation.
>
> The conclusion to be drawn from this is that the connection with the reality of the external world is established by the blind spot of the cognitive operation. *Reality is what one does not perceive when one perceives it.*[61]

The characterization of concepts in *What Is Philosophy?* and the definition of observation in Luhmann would thus seem to converge on a shared idea of the *production* of the outside as "not another site, but rather an off-site that"—like Luhmann's "environment," which is always already more complex than any system, and with which any system much achieve resonance if it is to remain operative—"erodes and dissolves all other sites."[62] "Like the structure of supplementarity whose logic it follows," Constantin Boundas writes—and like Luhmann's "environment"—"the outside is never exhausted; every attempt to capture it generates an excess or supplement, which in turn feeds anew the flows of deterritorialization and releases new lines of flight" ("Deleuze" 114–15).

At first glance, this seems strikingly similar to Luhmann's contention that reality is what one does not perceive when one perceives it, that "all observations have to presuppose *both* sides of the form they use as distinction or 'frame.' They cannot but operate (live, perceive, think, act, communicate) *within the world*. This means that something always has to be left unsaid, thereby providing a position from which to deconstruct what has been said."[63] But the inexhaustibility of the outside does not in Luhmann's account reside in the preexistent ontological fullness and positivity of the outside as such, as is often implied in Deleuze's work; instead, it resides in the possibility of other observations about x by other observers. For Luhmann, the paradoxical constitutive identity of x and y, inside and outside, is not unfolded by x's difference from itself, or from the environment's difference from itself, but rather by another observation, either by the same observer at a different point in time, or by a different observer at a different point in space—that is to say, not by the positivity of a generative environment but by the observing system making distinctions, and not simply by the system's observation, but by the observation of observation. For Luhmann, "x's difference from itself" is thus not a phenomenon of the outside but rather a production of the outside *from* the inside—that is, of an observation that is able to make meaning by "reentering" the distinction between x and y, inside and outside, on one side of the distinction itself, namely, the inside. After all, the distinction between language and not-language takes place within language; the distinction between figure and ground takes place within a frame.

This does not mean, however, that observation reconstitutes (to use Hegelian language) the identity of identity and nonidentity, because (as we saw earlier) the "the observation of observation" does not denote a Hegelian surpassing of a prior observation by a second observation that is its more total and fully realized Truth. The observations are linked differentially but not dialectically, not in a progressive movement toward more complete knowledge but rather in a circuit without a telos or center. There can thus be no "reflection" on and reconciliation of difference in the observation of observation, because such reflection on the constitutive "blind spot" of a given observation can only take place by means of *another* observation based on *another* distinction. Hence, even though there is radical equivalence between the "blind spots" of all observations, there is also radical, unbridgeable difference, and it is only on the basis of that difference, of a different constitutive and "blind" distinction and observation, that a critical view of any observed system can take place. Thus, the differences "between x and itself" that are in Deleuze created by the full positivity of being are in Luhmann produced by different observations; they are *discrete* and discontinuous. Observation will thus, in Luhmann's words,

> maintain the world as severed by distinctions, frames, and forms *and maintained by its severance....* This partiality precludes any possibility of representation of mimesis and any "holistic" theory. It is not sufficient to say that a part is able to express or to symbolize the whole....
>
> The operation of observing, therefore, includes the exclusion of the unobservable, including, moreover, the unobservable par excellence, observation itself, the observer-in-operation.[64]

What this means, then, is that for Luhmann—as he puts it in a phrase of nearly koan-like compression—"The world is observable *because* it is unobservable" ("Paradoxy" 46). "We resist the temptation," Luhmann playfully remarks, "to call this creation" (45).

Were we to leave matters here, we might be tempted to conclude that there is a difference of inflection only between Luhmann's observation and Deleuze's conceptualization, with Luhmann resisting the temptation to call observation "creation" (and in that resistance narrowly skirting philosophical ideal-

ism, not to say solipsism), and with Deleuze *not* resisting the temptation to call concept formation the "expression" (to use the term from his work on Spinoza) of objects
in the mind.[65] But a difference of more than inflection becomes clear with Deleuze's
development of the concept of "the fold" in the Foucault book and, later, in *The
Fold: Leibniz and the Baroque* (although the concept had been appearing sporadically
in his work, and in Foucault's, since the 1960s). As Deleuze introduces the concept
in *Foucault*:

> Up until now we have encountered three dimensions: the relations which
> have been formed or formalized along certain strata (Knowledge); the rela-
> tions between forces to be found at the level of the diagram (Power); and
> the relation with the outside, that absolute relation, as Blanchot says, which
> is also a non-relation (Thought). Does this mean that there is no inside?
> Foucault continually submits interiority to a radical critique. But is there *an
> inside that lies deeper than any internal world*, just as the outside is farther
> away than any external world? The outside is not a fixed limit but a moving
> matter animated by peristaltic movements, folds and foldings that together
> make up an inside: they are not something other than the outside, but pre-
> cisely the inside *of* the outside. (97)

Although the figure of the fold in *Foucault* is framed in terms of the problem of
"subjectivation," it is clear that in many ways it constitutes Deleuze's most ambitious attempt to refigure the problem of the relations between inside and outside,
the identity of the difference between both sides of the "cutting" that constitutes
Deleuzian conceptualization or Luhmannian observation. In the most general terms,
the figure of the fold theorizes a topographical relation between inside and outside
in which the existence of an individuated being depends, as Massumi puts it, "on a
constant infolding, or contraction, of an aleatory outside that it can only partially
control. The world is stable only to the extent that the strata working in concert
can regularize their infolding of chance; it is stable only within certain limits" (*User's
Guide* 53).

       This new topography enables, in turn, the theorization of sub-
jectivation within the frame of an attempted posthumanist ontology that Deleuze
situates explicitly against the phenomenological tradition of Husserl, Heidegger,
and Sartre. "Our point of departure," Deleuze writes, is "Foucault's break with phenomenology in the 'vulgar' sense of the term: with intentionality. The idea that
consciousness is directed towards the thing and gains significance in the world is
precisely what Foucault refuses to believe" (*Foucault* 108). The model for a type of
consciousness that is *not* "directed towards the thing," that is not "intentional" in

the strict phenomenological sense, is provided for Deleuze not only by Foucault but even more clearly, perhaps, by Spinoza, who shows that

> the body surpasses the knowledge that we have of it, *and that thought likewise surpasses the consciousness that we have of it.* . . . In short, the model of the body, according to Spinoza, does not imply any devaluation of thought in relation to extension, but, much more important, a devaluation of consciousness in relation to thought: "a discovery of the unconscious" — a non-Oedipal unconscious, of course — "of an *unconscious of thought* just as profound as *the unknown of the body.*"[66]

What Deleuze means here is that the objects of an aleatory outside impress themselves in the form of "ideas" upon the body, which infolds the effects of those objects in the form of thoughts — this is what it means to say "the body thinks" — and consciousness cannot fully capture or ever be totally aware of the body thinking. Consciousness, in other words, is partial and reductive, but it is also *not* a lack, as the Oedipal diagram would have it.[67] As Massumi explains it:

> The body infolds the *effect* of the impingement — it conserves the impingement minus the impinging thing, the impingement abstracted from the actual action that caused it and the actual context of that action. This is a first-order idea produced spontaneously by the body: the affection is immediately, spontaneously doubled by the repeatable trace of an encounter, the "form" of an encounter. . . . The autonomic tendency received second-hand from the body is raised to a higher power to become an activity of the mind. Mind and body are seen as two levels recapitulating the same image/expression event in different but parallel ways, ascending by degrees from the concrete to the incorporeal, holding to the same absent center of a now spectral — and potentialized — encounter. . . . This "origin" is never left behind, but doubles one like a shadow that is always almost perceived, and cannot but be perceived, in effect. ("Autonomy of Affect" 92–93)

Keeping in mind this picture of subjectivation as an infolding, we may now return to Deleuze's *Foucault* with a clearer idea of what Deleuze means when he writes that the subject of the fold is "never a projection of the interior; on the contrary, it is an interiorization of the outside. It is not a reproduction of the Same, but a repetition of the Different" (*Foucault* 98). "An Outside, more distant than any exterior," he continues,

> is "twisted," "folded," and "doubled" by an Inside that is deeper than any interior, and alone creates the possibility of the derived relation between

the interior and the exterior. It is even this twisting which defines "Flesh," beyond the body proper and its objects. In brief, the intentionality of being is surpassed by the fold of Being, Being as fold. (110)

This passage raises two crucial points, one that will enable us to zero in on the difference between Deleuze's "folded" concept of the inside/outside relation and that of systems theory, and the other which, raised by the prospect of "Flesh," will lead us directly into Alain Badiou's critique of the extension of the concept of the fold in Deleuze's book by the same name.

As for the first of these, it is clear, especially in light of the example of Deleuze's Spinoza, that what is being described in the "twisting," "folding," and "doubling" of Deleuze is a kind of "overcoding" (Massumi, *User's Guide* 51) or, better still, a "transcoding." In the example from Spinoza, the effect of the impingement is a transcoding of the stimulus from the outside (the body's "idea" of the impingement), which is then transcoded again by consciousness ("the idea of the idea"). The theoretical payoff for Deleuze here is obvious; he is thereby able to say that the thing, the stimulus, the event, is therefore *the same and not the same*. And this enables, in turn, an epistemological break with phenomenology; thought may no longer be seen as intention oriented toward an object, because thought is now retheorized as a nonlinear transformation, at each level of which the input is transcoded by a self-referential system that is selective according to its own rules. Thus, there can be no question of phenomenological transparency, and, moreover, the relation between different levels of transcoding is now revealed to be one of increasing complexity, transforms of transforms of transforms, fold upon fold upon fold. And insofar as those transcoding systems *are* self-referential (which they must be to transform and not merely *transmit* their input), they will apply their own rules to themselves recursively—a process that, as we have already seen, may lead to the emergence of new and unexpected states and forms.

From this vantage, the picture we get of the process of folding is thus one of fractal recursivity of the sort described by Francisco Varela. As Varela points out, the dynamic that is fundamental to the emergence of complexity out of fractal recursivity—and here we find a rather uncanny echo of Deleuze's figure—is one in which "operational closure generates a whole new domain in the apparently harmless act of *curling onto* itself."[68] In one well-known example, if we take a triangle, break each side of it in three to produce a six-pointed star, then take each side of the star and break these in the same fashion, and so on, what emerges from this recursive iteration of a simple rule—from this process of (un)folding—is the highly differentiated and extremely complex form of a snowflake, which is called a *fractal* because "the dimension of the final product is greater than 1 but less than 2"—a "fractional dimension" (316–17).[69]

But as similar as these figures of the recursive and fractal iteration of folding seem to be in Deleuze and systems theory, there is a crucial theoretical difference between them—a difference that will be only exacerbated, as we shall see in a moment, in Deleuze's development of the concept in *The Fold.* We may pinpoint the difference by reference to the examples from Deleuze that we have already examined. As Deleuze mobilizes it, the figure of folding as a transformation or overcoding depends on a relationship between inside and outside, system and environment, in which *information* or something very much like it is able to cross the line or "cut" of constitutive distinction. In the example from Deleuze's reading of Spinoza, there is *informational continuity* between the impingement, the effect of the impingement, the body's idea of that impingement, and consciousness's "idea of the idea"—that, after all, is what the "of" in the preceding phrase means. For systems theory, on the other hand, such "triggers" or "perturbations," as we have seen, carry no such information, so any talk of an "idea of an idea of an impingement" would be meaningless.

This may seem merely a quibble, but as Maturana and Varela argue, it is on this point that the break with the last vestiges of philosophical representationalism—and the difference between first-order and second-order cybernetics—rests. In direct contrast to the transformative infolding of the Spinozan body, Maturana and Varela argue that in a self-referential system, it is the "system's structural state that specifies what perturbations are possible and what changes trigger them."[70] What this means, in turn, is that the notion that the environment, the outside, contains information—*even if* we envision that information undergoing transcoding—is misleading (169). The problem with Deleuze, then, would seem to be that, on the one hand, he offers a perspective that is thoroughly constructivist and pragmatic, that insists on self-referential (and presumably nonrepresentationalist) coding; but, on the other hand, he smuggles representationalism back in in the "information-processing" model, in which information from the outside survives intact across the cut of distinction and directs the responses of a supposedly self-referential system, *even if* that information is increasingly diminished at each transformation.

From this vantage, the problem with the figure of the fold as Deleuze uses it is that it is not, after all, "twisted" in the manner described by Varela and Ranulph Glanville in "Your Inside Is Out and Your Outside Is In." As they argue, once it is acknowledged that observation is contingent—which is the assumption and indeed the imperative of the later Deleuze's view of concept formation—then it must also be acknowledged that constitutive distinction, because of its paradoxical identity, always turns back upon itself to "twist" and form a "strange loop," not simply a fold. Any putatively final distinction in either intension or extension will always generate another inside or outside, and so "at the extremes we find there are no extremes. The edges dissolve BECAUSE the forms are themselves continuous—they re-enter and loop around themselves" not like a circle, and not even like a fold, but like a Möbius strip.[71] What this means, in turn, is that because every observation is made by means of a "strange loop" of paradoxical distinction, "every world brought forth necessarily hides its origins. By existing, we generate cognitive 'blind spots' that can be cleared only through generating new blind spots in another domain" (Maturana and Varela, *Tree* 242).

So what the Deleuzian figure of the fold fails to take account of is that "as observers we can see a unity in *different* domains, depending on the distinctions we make." We can observe the internal states of a system, or we can consider how that system interacts with its "outside," its environment. For the first observation, "the environment does not exist"; for the latter, "the internal dynamics of that [system's] unity are irrelevant" (Maturana and Varela, *Tree* 135). But what is crucial is that

> both are necessary to complete our understanding of a unity. *It is the observer who correlates them from his outside perspective.* It is he who recognizes that the environment can trigger structural changes in it. It is he who recognizes that the environment does not specify or direct structural changes of a system. *The problem begins when we unknowingly go from one realm to the other and demand that the correspondence we establish between them (because we see these two realms simultaneously) be in fact a part of the operation of the unity.* (135–36; my emphasis)

We might say, then, that for Maturana, Varela, and Luhmann, the inside is *never*, strictly speaking, "the inside of the outside." Again, this is not merely a difference of inflection, because it captures nothing less than the difference between breaking with identity theory and sustaining it, the difference—to put it in suitably paradoxical Luhmannian terms—between theorizing the nonidentity, rather than the identity, of identity and nonidentity. As Luhmann puts it in "The Paradoxy of Observing Systems":

> Proceeding in this way from frame to frame or from form to form will, by necessity, reproduce the unmarked space. It will maintain the world as severed by distinctions, frames, and forms *and maintained by its severance.* "We may take it," to quote Spencer Brown, "that the world undoubtedly is itself (i.e. is indistinct from itself), but, in any attempt to see itself as an object, it must, equally undoubtedly, act so as to make itself distinct from, and therefore false to, itself. In this condition it will always partially elude itself." (44)

As long as information survives across the cut of constitutive distinction (as in Deleuze), difference will always be merely an epiphenomenon, however attenuated, of the identity of the effect or impingement from the outside that is continuous throughout the various levels and transcodings of the process. Systems theory, on the other hand, asks us to recognize instead that the outside is not anterior but is always produced "late," as it were; it is retroactively specified. More precisely, "observationally" we must see what the constitutive distinction is *first, before* we can see what is excluded by it—it is in this sense that the outside is the outside *of* the inside. Deleuze, on the other hand, goes "from one realm to the other" (as Maturana and Varela put it) and establishes a "topographical" correspondence between what are in fact "independent and nonintersecting phenomenal domains."[72]

This, I take it, is the point raised in a somewhat different tenor by Alain Badiou's reading of Deleuze's *The Fold*, which asks, in so many words, "How can a cut be a fold?" How can the "vacuum" between points be bridged? Badiou is sympathetic, of course, to the threefold aim of Deleuze's project: to found "an *antiextensional* concept of the multiple," "an *antidialectic* concept of the event," and an "*anti-Cartesian* (or anti-Lacanian) concept of the subject."[73] What Deleuze wants, Badiou writes, is

> absolute interiority, *but* "reversed" in such a way that it disposes of a relation to the All . . . a subject *directly* articulating the classical closure of the reflexive subject (but without reflexive clarity) and the baroque porosity of the empiricist subject (but without mechanical passivity). An intimacy spread over the entire world, a mind folded everywhere within the body: what a happy surprise! (61)

At the same time, however, Badiou resists what most distinguishes the figure of the fold in the Leibniz book from its presentation in *Foucault;* in the latter, we find only a fleeting gesture toward the "peristaltic" quality of the fold, but here, Badiou notes, we have a "vision of the world as an intricate, folded, and inseparable totality such that any distinction is simply a local operation," a view of the "multiple as a large animal made up of animals, the organic respiration inherent to one's own organicity, the multiple as *living tissue*, which folds as if under the effect of its organic expandings and contractings" (55).

Instead of the essentially abstract, fractal, and recursive version of the fold that we get in *Foucault*, what Deleuze offers us here is a radical philosophical expressionism in which singularity is but a momentary burp or hiccup in the substantial body of the world in peristaltic movement. As Badiou points out, if we believe Deleuze's assertion that "there is no vacuum between two points of view" (quoted in Badiou 63), then "this absence of a vacuum introduces a complete continuity between the points of view." Thus, "ontological organicism forecloses the vacuum"—a vacuum created in systems theory by the cut of constitutive distinction— "according to the law (or desire, it is the same thing) of the Great Animal Totality" (63). What Deleuze gives us is thereby "a philosophy 'of' nature, or rather a philosophy as nature. This can be understood as a *description in thought of the life of the world*, such that the life thus described might include, as one of its living gestures, the description itself" (63). If systems theory courts solipsistic idealism—as Luhmann in so many words acknowledges when he admits that "we resist the temptation to call [constitutive distinction] creation"—then *The Fold* seems to embrace the sort of holism we have already seen in the late Gregory Bateson, who thought that "the individual mind is immanent but not only in the body. It is immanent also in pathways and messages outside the body; and there is a larger Mind of which the individual mind is only a subsystem. This larger Mind is comparable to God and is perhaps what some people mean by God."[74]

Deleuze's vision of *The Fold* complexifies this account, to be sure, but it does not finally break with it; the pathways and networks may be infinitely and fractally infolded, but the relationship of immanence and continuity between subfolds and the totality remains intact. And Deleuze's substantialization, his "ontological organicism," as Badiou puts it, only exacerbates the problem of how we can continue to hold, as Boundas does, that Deleuze conceives the outside as a reservoir produced by the logic of supplementarity. As Badiou argues: "That there be excess (indifferently shadow or light) in the occurrence of the event, that it be creative, I agree. But my distribution of this excess is opposed to Deleuze's, who finds in it the inexhaustible fullness of the world" (65). "For me," he continues—and here he would seem to be in agreement with Luhmann, for whom the outside always lies on the other side of the "cut" of distinction, always comes *after* and always in a sense too *late*—

it is not from the world, even ideally, that the event gets its inexhaustible reserve, its silent (or indiscernible) excess, but *from its not being attached to it.* . . . The excess of the event is never related to the situation as an organic "dark background," but as a multiple, so that the event *is not counted for one by it.* The result is that its silent or subtracted part is an infinity *to come,* a postexistence that will bring back to the world the pure separated point of the supplement produced by the event. . . . Where Deleuze sees a "manner" of being, I say that the worldly postexistence of a truth signals the event as *separation.* (65)

For Deleuze, this is the view characteristic of much modern philosophy, in which "bifurcations, divergences, incompossibilities, and discord belong to the same motley world *that can no longer be included in expressive units*" (as they can in the world of Leibniz's fold).[75] The price we are asked to pay to gain that expressivity and univocity, however, is a high one indeed, because, as Boundas points out, "Leibniz may have been the grand theorist of the event"—of "the triumph of the wave over the particle and the fold over the cut or vacuum"[76]—

but he never failed to be also the grand advocate of god: the principle of sufficient reason, placed by him in the service of the theological, reassuring discourse of the "best possible world," subjected divergence and disjunction to a negative use. His individual/points of view come to be and to form a series only insofar as they all converge upon the same town. (Boundas 109)[77]

In the end, though, as Badiou points out, Deleuze has recourse to one final strategy in *The Fold,* a strategy less transcendental and more pragmatic, for negotiating the problems we have been discussing, and that is to regard philosophy—to use the language of speech-act theory—not as constative but as performative, as a kind of *writing* that, so to speak, does the impossible. *The Fold*'s vitalism and organicism may compromise Deleuze's confrontation with the paradoxes of the outside by attempting to suture closed the vacuum between points (or observations) with Total Substance. But its writing practice *performs* the complexities of the distinction in a fashion perhaps unparalleled by *Foucault,* and in so doing traffics in the gap, as it were, between the two sides of the strange loop that Deleuze cannot bear to keep strange. This philosophical writing practice, this performative shuttling, as Badiou puts it,

marks a position of hostility (subjective or enunciating) with respect to the ideal theme of the clear, which we find from Plato (the Idea-as-sun) to Descartes (the clear Idea), and which is also the metaphor of a concept of the Multiple that demands that the elements composing it can be exposed, by right, to thought in full light of the distinctiveness of their belonging.... Nuance is here the antidialectic operator par excellence. Nuance will be used to *dissolve* the latent opposition, one of whose terms the clear magnifies. Continuity can then be established locally as an exchange of values at each real point, so that the couple clear/obscure can no longer be separated, and even less be brought under a hierarchical scheme, except at the price of a global abstraction. This abstraction is itself foreign to the life of the world. (54)

If we believe Badiou, then, Deleuze's philosophical pragmatism is not so much surrendered in *The Fold* as it is relocated, always at work even in his most resolutely metaphysical moments. And in that light, the pragmatism of the late Deleuze is not so much a problem of philosophy as it is a *mode* of philosophy—a pragmatism of philosophical practice of the sort that Deleuze found in Foucault, who

is not content to say that we must rethink certain notions; he does not even say it; he just does it, and in this way proposes new co-ordinates for praxis. In the background a battle begins to brew, with its local tactics and overall strategies which advance not by totalizing but by relaying, connecting, converging and prolonging. The question ultimately is: *What is to be done?* (*Foucault* 30)

Conclusion

Post-Marxism, Critical Politics, and the Environment of Theory

THROUGHOUT THE preceding chapters, I frequently have made recourse to how a post-Marxist critical perspective can reveal the political limits of many of the theoretical paradigms I have examined thus far. As we have seen, the work of Rorty, Cavell, Michaels, Luhmann, and Maturana and Varela (among others) runs aground time and again on the ideological recontainment of a potentially liberating epistemological and philosophical pluralism by a pluralism of a very different and more familiar sort—a *liberal humanist* pluralism (or something very much like it) that pays little attention to how real inequality in the economic and social sphere complicates and compromises the pluralism they imagine generated by their constructivism in the sphere of theory. The most heavily canonized version of this charge probably remains the classic Marxist critique of the liberal concept of rights, which holds that the supposition of "abstract" equal rights under the law in civil society masks a more fundamental inequality in economic power, so that the reality of "equal" human rights turns out to be *unequally* distributed *property* rights. Even in the politically attuned poststructuralism of Foucault and Deleuze, the specificity of this problem—and of what Louis Althusser famously calls the "overdetermination" of the social totality by the "structural causality" of the capitalist mode of production—is largely if not totally sidestepped:[1] for Foucault, in favor of an analytics of power whose links to the systematic control of wealth are woefully undertheorized, for reasons having not least of all to do, as Abdul JanMohamed has pointed out, with Foucault's fretful "disavowal" of Marx, which is in turn "symptomatic of a hasty 'post-Marxism' that has never adequately come to terms with Marx."[2] And for Deleuze (and Guattari), the lack of sustained attention to the problem of structural causality is ironic indeed given their near fetishization of the new, the unthought, and the opening of "lines of flight," all of which bears more than a passing resemblance to the fetishization of difference and the new *tout court* in postmodern commodity culture generally.[3]

In light of these reservations, it is only fitting, then, that I conclude this study by moving to consider post-Marxist theory in its turn, and I want to do so by way of a very specific question: If we find the theoretical arguments for the "constructivist turn" in theory examined in these pages persuasive, then what difference does that constructivism make to one's view of the relationship between theory and politics? To answer that question as specifically and clearly as I can, I want to examine the work of the most important Marxist theorist of the past two decades in America, and surely one of the most influential theorists generally of our day — Fredric Jameson — not least of all because Jameson (unlike post-Marxists such as Ernesto Laclau and Chantal Mouffe) has maintained a firm commitment to a "totalizing" critique while at the same time attempting to accommodate and incorporate much of the best work in constructivist postmodern theory. And to give an even more finely grained sense of how my view of the relation of theory to politics differs from Jameson's, I want to punctuate my discussion with reference to two recent exemplary discussions of Jameson within the field of postmodern theory generally, in Stephen Best and Douglas Kellner's *Postmodern Theory: Critical Interrogations*, and in Barry Smart's *Modern Conditions, Postmodern Controversies*.

As I hope will be clear, I share a commitment to many if not all of the political aims expressed in both books: to Smart's call for "a regeneration of democratic politics as the sole vehicle for implementation of the values of liberty, diversity, tolerance, and solidarity which alone offer 'a chance of a better society,'"[4] but qualified by the rather more Marxist proviso of Best and Kellner that what is needed to provide a "concrete and substantive basis" for "a radical political alliance" that can make good on these democratic values is "a common anti-capitalist politics" (292). So my disagreement with both studies is not primarily in political values, but rather in how the *relation* between those values and the work of theory is to be construed. It is that difference-in-identity, I hope, which will provide a clear sense of my own position on the full implications of the constructivist challenge for politically engaged theory.

In light of the problem that has framed this study—the problem of how different theories account for their own "outside"—perhaps the best place to begin our discussion of Jameson is to realize that for him, as Smart puts it, "the effects of the postmodern condition"—famously characterized by Jameson as above all the experience of a "new depthlessness" in social life, and a subsequent loss of the ability "to map the great global multinational and decentered communicational network in which we find ourselves caught as individual subjects" (quoted in Smart 186, 187)—"are not considered to extend to totalising forms of analysis, or emancipatory narratives on the subject of socialism. In consequence, as the 'abolition of critical distance' and the process of mutation of the cultural sphere under late capitalism is outlined, the impression is simultaneously conveyed of the existence of another space or place, outside and beyond the sphere of influence described, from which, by implication, a privileged analysis can continue to be conducted" (187). For Jameson, that "space or place" is, of course, the theoretical locus of totalizing Marxist analysis itself, a totalization that makes possible Jameson's well-known analysis of postmodernism as the "cultural dominant" of a third phase of capitalist development—"late" or multinational capitalism—that follows the earlier modes of monopoly capitalism (with its cultural dominant of modernism) and, before that, market capitalism (with its cultural dominant of realism). As Best and Kellner explain, totalization is crucial in Jameson's view to any genuinely critical analysis of society and culture that hopes to be anything other than a mere inventory of stylistic features, and he defends it on two main counts: "(1) difference itself cannot be genuinely understood outside of a relational and systemic context; (2) a totalizing analysis is necessary to map the homogenizing and systemic effects of capitalism itself" (Best and Kellner, *Postmodern Theory* 187).

The question immediately raised here, of course, is how it is possible to have a "genuine" understanding and authoritative "map" of the play of postmodern differences and their effects *without at the same time* being subject to that play. How, indeed, to diagnose the utter eclipse of critical distance under postmodernism and at the same time hold that there is a critical perspective from which a definitive social cartography can be undertaken? Jameson's response to this dilemma, in a variety of contexts, has been that Marxist analysis, while authoritative, is not exactly "transcendent" in the usual sense. This is perhaps clearest in his critique (in *Postmodernism, or, the Cultural Logic of Late Capitalism*) of Walter Benn Michaels's *The Gold Standard and the Logic of Naturalism*, which would seem to present us with an inescapable dilemma: either step outside the social system configured by the logic of capitalism (on pain of indulging metaphysics, theology, or the "theory" of "Against

Theory") and imagine the possibility of a real alternative to it, or stay locked within the system and its logic (the only thing one can do, it would seem, in good postmodern faith) and reproduce a system that thrives on irreducible constitutive difference (simply because such difference is the structure of desire that both fuels and is fueled by commodity culture), however much you may think you want to do otherwise.

Jameson's response to the immanence/transcendence problem modeled in Michaels's study is that one need not be able to step outside the social system into some metaphysical or transcendent space to think the possibility of social and historical change—or to accomplish a "genuine" cartography of the present. For Jameson, the possibility of social change resides *within* the existing social system itself (where else would it be?), because the economic and social totality is never self-identical and uniformly dispersed, but is always internally differential and discontinuous—the prime example of this being the unequal relations of production at work in the fact of *class*. According to Jameson, Marx's aim in *Capital* is to demonstrate that the alternative to the capitalist totality is to be found precisely *within* the logics and dynamics of capitalism itself—not, in Jameson's words, "as an ideal or a Utopia but a tendential and emergent set of already existing structures."[5] As Marx and Engels put it at the end of the first section of *The Communist Manifesto*, "What the bourgeoisie, therefore, produces, above all, are its own gravediggers." For Jameson, this is the "strong" form of what Marx means by the concept of "contradiction"—most centrally in the classical Marxist canon, of course, the contradiction between the development of the forces and relations of production, which is in turn the very engine of class formation and of historical change itself.

It is this internally generative nature of contradiction that is accentuated in Louis Althusser's influential theorization of the relationship between contradiction and "overdetermination." For Althusser, Marxist analysis "always studies economic structures dominated by *several* modes of production,"[6] and it is that "unevenness" or overdetermination which is "the motor of all development" (*For Marx* 217). For Althusser and for Jameson (and for all Marxists), however, the theoretical problem is how to retain this strong sense of contradiction—how, to use Jameson's formulation in *The Seeds of Time*, to keep contradiction from degenerating, as it were, into "mere antinomy"[7]—without at the same time falling into the false assurance of some notion of teleological inevitability. And even if the notion of "teleological assurance" reputed to Marxist theory is indeed, as Jameson has suggested, something of "an ideological straw- or bogey-man,"[8] the explanatory priority of the economic as the determining engine of social change surely is not.

What is at stake here, in other words, is nothing other than the status of Marxism's base-superstructure model itself. Although post-Marxists such as Ernesto Laclau and Chantal Mouffe clearly reject that model as such,[9] it was pursued to exhaustion within the Marxist tradition itself in the work of Althusser, who attempted to solve the problem by insisting on a reciprocal relationship between economic base and "semiautonomous" superstructure, *while at the same time* conceding the determination of superstructural forms by the economic mode of production, but only "in the last instance"—only to famously admit later that "the last instance never comes."[10] The negotiation of this impasse was the aim, of course, of Althusser's concept of "structural causality," which attempted to replace the crude economic determinism harbored by the base-superstructure model with a more syncretic model of totality in which, as Althusser defines it, "*the whole existence of the structure consists in its effects*...is merely a specific combination of its particular elements, is nothing outside its effects" (*Reading Capital* 189).[11]

I have gone on at some length about Althusser's intervention here because it is crucial to Jameson's conjugation of the relation between politics and theory on the terrain of the postmodern. To begin with, Althusser's model of structural causality is the most influential precursor within Marxist theory for Jameson's attempt to reconcile the priority of Marxist theory—and therefore the priority of the economic within the domain of the social—with sufficient justice to the effects of postmodern difference and depthlessness. This is very much the register in which we are meant to take Jameson's assertion that postmodernism—exemplified *aesthetically* in the video art of Nam June Paik, *technologically* in the computer, and *architecturally* in the Westin Bonaventure Hotel in Los Angeles—is the cultural dominant of a "vaster and properly unrepresentable totality which is the ensemble of society's structures as a whole" (*Postmodernism* 51). And those structures are in turn themselves "but a distorted figuration of something even deeper"—and here Jameson turns toward Lukács's "expressive" rather than Althusser's "structural" causality—"namely the whole world system of a present-day multinational capitalism" (37).[12] It is of signal importance to underscore this second, more Lukácsian turn of Jameson's theory, for it enables him to assimilate the various forms of postmodern *theory* to his totalizing model as well; "what is today called contemporary theory—or better still, theoretical discourse—" Jameson writes, "is also, I want to argue, itself very precisely a postmodernist phenomenon" (*Postmodernism* 12). So it is, according to Jameson, that "every position on postmodernism in culture—whether apologia or stigmatization—is also at one and the same time, and *necessarily*, an implicitly or explicitly political stance on the nature of multinational capitalism to-day" (3).

Two crucial issues are raised by this attempt to outflank the various forms of postmodern theory by Marxist totalization: first, from what ground or vantage can one know all of this?; and second, has Jameson here run afoul of his own quite appropriate admonition about the inadequacy of taking *moral* positions on postmodernism? As for the first question, Jameson provides the answer that has typically served as the Marxist tradition's fallback position (though one senses at times that it strikes him as a somewhat anachronistic and uncomfortable one): that "space or place" of theoretical authority is provided by Marxist "science," the epistemologically secure and politically efficacious opposite number to "ideology" familiar to us since (at least) Marx and Engels's *The German Ideology*. As Jameson forthrightly admits early on in *Late Marxism*, " 'To be a Marxist' necessarily includes the belief that Marxism is somehow a science: that is to say, an axiomatic, an organon, a body of distinctive knowledges and procedures."[13] It is "Marxist science" that provides a space where one can achieve the "critical distance" on the postmodern play of difference that Jameson tells us is available nowhere else in postmodern society. And it is Marxist science that allows an authoritative assimilation of the heterogeneities of postmodern culture—including the varieties of postmodern theory—to a model in which they are essentially symptomatic of a primary economic determination. As Jameson puts it at the end of his most important essay on postmodernism:

> Althusser's formulation remobilizes an older and henceforth classical Marxian distinction between science and ideology that is not without value for us even today. The existential—the positioning of the individual subject, the experience of daily life, the monadic "point of view" on the world to which we are necessarily, as biological subjects, restricted—is in Althusser's formula implicitly opposed to the realm of abstract knowledge.... What is affirmed is not that we cannot know the world and its totality in some abstract or "scientific" way. Marxian "science" provides just such a way of knowing and conceptualizing the world abstractly.... The Althusserian formula, in other words, designates a gap, a rift, between existential experience and scientific knowledge. Ideology then has the function of somehow inventing a way of articulating those two distinct dimensions with each other. (*Postmodernism* 53)

It is here, I think, that the Althusserian legacy—initially so promising in its attempt to think irreducible social complexity through structural causality—exacts a price we should not be willing to pay; for the fee it levies in clearing an epistemologically privileged space above the social fray of difference, from which the priority of the economic can be asserted, is to thrust us, with Jameson, into the very philosophical position he finds wanting in the different varieties of postmodern "formalism" and "neo-Kantianism." That price, of course, is philosophical *idealism*—and idealism very much in the sense that Adorno himself had in mind: a kind of violence and rage against the materiality and heterogeneity of what Adorno called "the preponderance of the object," whose nonidentity and difference the "concept" of idealism and "identity theory" attempts to suppress and master in thought.[14] We need to remember here that Jameson's central charge against the antifoundationalism of Rorty and other postmodern theorists is its "formalist impulse": that, in refusing the challenge to theorize "a certain harmony" between "a specific social content" and "group or collective structures" that "must emerge from concrete historical development" (*Seeds* 44), it seeks to achieve "formal purity" by abandoning the problem of historical embeddedness altogether. But that sort of formalism—or what Althusser in his later work acknowledges as his earlier "theoreticism"[15]—is precisely what is at work (as Marxist commentators such Sebastiano Timpanaro have noted) in the "epistemological idealism" of the Althusserian notion of "science," a fact that is *especially* clear when we remember that, for Althusser, science is (as he notoriously put it) a "subjectless" procedure.[16]

It may initially seem strange to characterize this view as "idealism"—rather than, say, realism, materialism, or objectivism—but, as Martin Jay points out in his encyclopedic discussion in *Marxism and Totality*, whereas "Orthodox Marxism generally wavered between traditional scientific realism based on a correspondence theory of truth (sometimes, to be sure, understood asymptotically) and a pragmatist notion of verification as the historical realization of predictions," Althusser's concept of "science" rejects all of these. Instead, it sees the truth of science as residing "entirely within the dialectical logic of the concept" through "an epistemology of conceptual self-correction" (399). Thus, "truth is the sign of itself and not verifiable by any external criterion"—a position that relies on "a large measure of faith and circular reasoning" (401), and invites Paul Ricoeur's characterization of structuralist "science" as "Kantianism without a transcendental subject" (quoted in Jay, *Marxism and Totality* 389). The "realism" of Althusser's Marxist "science," in other words, is possible only on the basis of a prior epistemological idealism that untethers it from the play of differences and contingencies thought to be otherwise inescapable in the postmodern social field.

It is certainly true, as Jameson reminds us, that we misunderstand the category of totality—as some of the more facile celebrations of difference in postmodern theory do—if we assume "that philosophical emphasis on the indispensability of this category amounts...to a celebration of it" (*Late Marxism* 27). But if Marxist science does what Jameson says it does, then celebration is very much beside the point anyway, in which case we must disagree with Jameson's assertion that we misunderstand the philosophical stress on totality if we think that it amounts to an "implicit perpetuation [of totality] as a reality or referent outside the philosophical realm" (27). For that, indeed, would seem to be precisely the status accorded to totality by Marxist science, insofar as the economic mode of production is just such a "reality or referent" that totalizes the social field via causality (either Lukácsian "expressive" or Althusserian "structural").

This much is clear, it seems to me, in Jameson's comments on totalization in the conclusion to the postmodernism book, which are worth quoting at some length:

> If we object that the philosophical dilemma or antinomy hereby evoked holds only for absolute change (or revolution), and that these problems disappear when the sights are lowered to punctual reforms and to the daily struggles of what we may metaphysically call a kind of local politics (where systematic perspectives no longer hold), we have of course located the crucial issue in the politics of the postmodern as well as the ultimate stake in the "totalization" debate. An older politics sought to coordinate local and global struggles, so to speak, and to endow the immediate local occasion for struggle with an allegorical value, namely that of representing the overall struggle itself and incarnating it a here-and-now thereby transfigured. Politics works only when these two levels can be coordinated; they otherwise drift apart into a disembodied and easily bureaucratized abstract struggle for and around the state, on the one hand, and a properly interminable series of neighborhood issues on the other, whose "bad infinity" comes, in postmodernism, where it is the only form of politics left...a situation in which, for a time, genuine (or "totalizing") politics is no longer possible; it is necessary to add that what is lost in its absence, the global dimension, is very precisely the dimension of economics itself, or of the system, of private enterprise and the profit motive, which cannot be challenged on a local level. I believe that, *en attendant*, it will be politically productive, and will remain a modest form of genuine politics in its own right, to attend vigilantly to just such symptoms as the waning of the visibility of that global dimension, to the ideological resistance to the concept of totality, and to that epistemological razor of postmodern nominalism which shears away such apparent abstrac-

tions as the economic system and the social totality themselves, such that for an anticipation of the "concrete" is substituted the "merely particular," eclipsing the "general" (in the form of the mode of production itself). (*Postmodernism* 330)

There are many things to be remarked in this passage, not the least of which is the assertion that "politics works only when these two levels can be coordinated," and that otherwise resistance is "disembodied and easily bureaucratized." Jameson is certainly right to stress what he sees as the central work of theoretical totalization: "to show that no 'philosophical concept' is adequate either: each one must be analyzed symptomatically for what it excludes or cannot say" (*Late Marxism* 37).

What I want to argue now, however, is that this task may be accomplished without the belief that, perforce, "those concepts demand dialectical analysis" (37). As we have seen, the theory of the observation of observation in systems theory accomplishes just such a task, but *without* the idealism (of Marxist "science") and the reductionism (of the social to the economic as the totalizing immanent cause of postmodern difference) that accompanies the Marxist dialectic. To hold as much is, to be sure, to surrender the authority of Marxist science. But is it thereby—as Jameson suggests—to surrender any claim to political praxis? It might well be argued instead, as Smart responds, "that political action and struggle has not ceased or been lost, but rather has increasingly assumed 'inconvenient' non-class forms" (190)—the very forms of the well-known "new social movements" such as feminism, gay and lesbian rights, environmentalism, animal rights, and so on.

Indeed, as we have already seen, here is where the interventions of Foucault and Deleuze prove invaluable for a pragmatist politics of resistance in their analytics of institutions and of the productivity of power, their insistence on the importance of investment and its relation to material reproduction for social dynamics, and their recognition of the full range of "forces" at work on the micropolitical level of society. Such analysis is crucial to "the new social movements" in disclosing, for example, how the mechanisms of discipline are inscribed on the subject's very body, in her very actions, through regimes of "health," sexuality, and other "technologies of the self," which in traditional Marxism are discounted as "epiphenomenal" or "diversionary" sites of social struggle. Similarly, Deleuze's work on how minoritarian sexual forms open up the body and its forces as the "outside" of liberal society goes to the very heart of the political under postmodernism, because it challenges the very form of subjectivity on which liberal democracy relies— the diagrammatic regime of Oedipal normalization—a fact whose political stakes, as Brian Massumi reminded us in the preceding chapter, have been much more acutely recognized by the New Right than the Left, and with devastating political effect.

What is interesting about moments such as these in Jameson's work is that, as Best and Kellner point out, "there is a tension in Jameson's writings, theoretically, between the privileging of Marxism as the master discourse and the perspectivism of standpoint theory. Politically," they continue, "there is a tension between a traditional class politics and a more pluralist alliance politics." But "whatever position Jameson upholds," they conclude (and here they would seem to agree with Smart),

> he has not established that the complexification and fragmentation of "the working class" under postwar and postindustrial conditions does not inalterably change the composition of class relations and politics. Any further clarification of his position should state how the "proletariat" can be expected to become a unified subject again (if indeed it ever was) and why it should remain the epicentre of political struggle. (191)

It is probably clear by now that I agree with Best and Kellner on this point, but I disagree with them about how that characterization affects one's recourse to the dialectic as they discuss it in the last two chapters of their study:

> We would argue that a dialectical social theory such as one finds in the best of critical theory [and here they mean specifically the Frankfurt School and its inheritors] provides the most adequate models and methods to analyze the multidimensional processes toward both fragmentation and unification, implosion and differentiation, and plurality and homogenization in contemporary techno-capitalist societies. . . . In contrast to the postmodern caricature of dialectics as a mystical and teleological logic of history, dialectics for critical theory is primarily a method for describing relationships between different domains of social reality, such as the economy and state or culture. (224)

This characterization might be persuasive were it not for the fact that the example of Jameson—our foremost contemporary practitioner of the dialectic, it must be admitted—makes it clear (in passages such as the one quoted earlier from *Postmodernism*) that embracing the dialectic commits us to a good deal more than this. As Jameson points out, "anti-Utopian" postmodern critics charge the dialectic and totalization with positing

> the end or master term of all such themes as this or that variant of a still essentially Hegelian notion of "reconciliation" . . . which is to say, the illusion

of the possibility of some ultimate reunion between a subject and an object.... "Reconciliation" in this sense, then, becomes assimilated to this or that illusion or metaphysic of "presence," or its equivalent in other postcontemporary philosophical codes. (*Postmodernism* 334–35)

This constitutes the familiar postmodern strategy of reproaching Marxism "with its temporal dimension, which allows it to consign solutions to philosophical problems to a future order of things," even if—as in the case of Adorno's "negative dialectics"— that future reconciliation is (paradoxically) infinitely deferred, under pain of reification and the tyranny of identity and the concept (Adorno, *Negative Dialectics* 231).

But there is a crucial difference between the *dialectical* means of deferral (the escape hatch through which, as Adorno well realized, the dialectic is able to skirt charges of Hegelian reconciliation of subject and object) and the sort of deferral that emerges from the temporalization of paradox and antinomy in systems theory. And this difference has to do with the Marxist recourse to the self-presence of "science." To sharpen our sense of what is at stake here, we need to understand that Jameson's point is not *simply* that all interpretations are reductive of the verticality of difference, and that therefore one should then be reductive in the direction of emphasizing the priority of the economic; for that would constitute an essentially *pragmatist* defense of the position, a defense that Jameson is clearly at pains to reject. Nor is his point simply that dialectical totalization allows "a mapping of structural limits for which causality must rather be redefined as the *conditions of possibility*" (*Seeds* xv). Rather, Jameson's point—it is the "strong" reading of the need for dialectics, as he would say—is that there is *one and only one* vantage from which those "structural limits" can be *authoritatively* mapped, one and only one vantage from which all other interpretations can be revealed as reductive and thereby assimilated as "symptoms" of postmodernism: the dialectical perspective of Marxist science and totalization.

What the dialectic can only struggle with under the guise of the deferral of Hegelian reconciliation (an altogether necessary deferral, if the dialectic is to escape the charges of identitarian and teleological thinking leveled against it by poststructuralism) is theorized more powerfully and productively by systems theory and its related strands of poststructuralism. This is so because the "conditions of possibility" identified by them are also, as Rodolphe Gasché has put it, conditions of *impossibility*, insofar as the identification of those conditions will always be a contingent interpretation that could have been otherwise (and an interpretation, moreover, constructed atop a constitutive blind spot of paradoxical distinction that can only be disclosed by—deferred to, if you will—a *second-order* observation).

This is why Jameson is incorrect, I think, in asserting that "the very concept of differentiation . . . is itself a systematic one; or, if you prefer, it turns the play of differences into a new kind of identity on a more abstract level" (*Postmodernism* 342). Precisely the opposite would appear to be true. Properly understood, the concept of differentiation as Luhmann uses it turns those differences, as we have seen, into a new kind of *non*identity, in that systems by definition remain blind to their paradoxical, constitutive distinctions, which only *other* observing systems can see. Hence, differential systems do not coalesce vertically into a single identity at another, "more abstract level," but remain horizontally dispersed in the social field.

In this light, we may now identify the antinomy (or contradiction, if one prefers!) between Best and Kellner's commitment to the dialectic and their call "for theory to be reflexive and self-critical, aware of its presuppositions, interests, and limitations," a theory "non-dogmatic and open to disconfirmation and revision, eschewing the quest for certainty," one that is "non-scientistic, fallibilistic, hermeneutical, and open to new historical conditions, theoretical perspectives, and political applications" (257–58); for what is *not* theorized by the dialectic—by that Marxist totalization which allows us to assimilate all the positions in the postmodern field as expressions of multinational capitalism—is what systems theory more rigorously pursues, and what Best and Kellner say they want: a "multiperspectival social theory" that emphasizes that each interpretive act, each social description, is "a way of seeing, a vantage point or optic to analyze specific phenomena" (264–65).

Now, having said as much, I hasten to add that I *agree* with Best and Kellner that what is needed at the current moment is a "common anti-capitalist politics" among different alliance groups; I agree that the "exploitation and repression of diverse groups and individuals in the capitalist economy and state provides a fundamental point of commonality to unite a myriad of oppressed social groups" (292). What I do not agree with is their view that this pressing political project requires—or is in fact even theoretically compatible with—dialectical totalization. The case for a common "anticapitalist" politics, if it is to be made, must be made on *pragmatic* and not foundational grounds, must be raised on the sort of framework provided by Ernesto Laclau and Chantal Mouffe's post-Marxism, which holds that "progressive values must be defended within a pragmatic context that appeals to the non-arbitrary force of sound argumentation and discursive strategies" (Best and Kellner 198).

This sort of pragmatic defense of the need for a common anticapitalist politics is very much in keeping with the "multiperspectival" kind of theory

Best and Kellner say they want, but it is a pragmatism rescued from its often-in-dulged liberal complacency by two crucial commitments: first, a ceaseless engage-ment with, rather than "evasion" of, epistemology-centered theory and the broader questions of interpretive contingency and difference that it raises; and second, re-newed attention to the centrality of the problem of *capitalism* (and not just "democ-racy," as in Richard Rorty).

As for the first of these, systems theory extrapolates, rather than short-circuits, the pragmatist commitment to contingency as the hallmark of a gen-uinely pluralist philosophy. Luhmann, like Rorty, stresses the contingency of inter-pretation and observation but—contra Rorty—derives from their paradoxicality the *necessity* of the observations of others, thus pluralizing the Rortyan "we." For it is only in the distributed observations of *different* observers that a critical view of any observed system, or any social fact, can be constructed. And, as I have argued, although this reformulation is neither an ethics nor a politics per se, it *does* provide a rigorous and compelling theorization of the conditions of possibility for pluralist *sociality as such*. As we have seen, Luhmann's insistence on the "blind spot" of obser-vation and, therefore, on the essential aporia of any authority that derives from it (the authority, say, of the system that enforces the distinction legal/illegal) may be fruitfully paralleled with the theory of democratic "social antagonism" in the work of Ernesto Laclau, Chantal Mouffe, and Slavoj Žižek. These theorists, like Luh-mann, do not disavow or repress the "broken and perverted" (i.e., paradoxical and tautological) nature of communication, but rather derive from it the conditions of possibility for democratic sociality. As Žižek puts it, "the limitation proper to the symbolic field as such" is "the fact that the signifying field is always structured around a certain fundamental deadlock,"[17] or what Luhmann characterizes as the "block-age" of paradoxical self-reference. Like the theorists of social antagonism, Luh-mann insists that the distribution or unfolding of such "blockages" or paradoxes is not an impediment to democratic society but is in fact *crucial* to it. And hence, a truly pluralist, "multiperspectival" philosophy should avoid at all costs the quintes-sentially modernist and Enlightenment strategy of reducing complexity via stable social consensus.

And here, precisely, is where the systems theory renovation of pragmatism is joined not only by the theory of social antagonism, but also by the work of Deleuze, which provides, as it were, the ontology, rather than the epistemol-ogy, for the conditions of possibility for democratic pluralism. As in systems theory's commitment to irreducible complexity and the distribution of observation in a hor-izontal, functionally differentiated social space, Deleuze's work, as Michael Hardt suggests, helps us "develop a dynamic conception of democratic society as open, horizontal, and collective." "This open organization of society," Hardt continues, "must be distinguished from the vertical structures of order"; it is not a "plan or blueprint of how social relationships will be structured," but rather "a continual process of composition and decomposition through social encounters, on an immanent field of forces."[18]

As for the second imperative — the necessity of an analysis of *capitalism* as such and not just of democracy — the pragmatic commitment to an anticapitalist politics is as crucial to systems theory as it is to mainline liberal pragmatism, for it can confront systems theory as well with its own worst tendencies. As I have argued in my accounts of Maturana, Varela, and Luhmann, those tendencies more often than not involve indulging its own brand of idealism — a brand that links it to the idealism of both Rorty and Cavell — in failing to account for the inequities of power that complicate and compromise the formal equivalence of different observers in the social field. We need to remember — as Jameson puts it in what we hope would serve as a reminder to Luhmann — that "no matter how desirable this postmodern philosophical free play may be, it cannot now be practiced; however conceivable and imaginable it may have become as a philosophical aesthetic (but it would be important to ask what the historical preconditions for the very conception of this ideal and the possibility of imagining it are), anti-systematic writing today is condemned to remain within the 'system'" of global capitalism and the law of value (*Late Marxism* 27).

What I have in mind here, then, is something like what Kenneth Burke calls a "comic" perspective on the relationship between Marxist theory and a pragmatism renovated by systems theory, between totalization and detotalization, dialectic and antinomy, the assertion of material, social realities and the self-critical constructivist epistemology that should always interrogate those assertions. As Burke puts it, a comic frame shows us "how an act can 'dialectically'" — and I note the scare quotation marks — "contain both transcendental and material ingredients, both imagination and bureaucratic embodiment. . . . It also makes us sensitive to the point at which one of these ingredients becomes hypertrophied, with the corresponding atrophy of the other."[19] The comic frame, he continues, should thus "enable people *to be observers of themselves, while acting.* Its ultimate would not be *passiveness,* but *maximum consciousness.* One would 'transcend' himself by noting his own foibles" (171; emphasis in the original). The comic frame, Burke concludes, "considers human life as a project in 'composition,' where the poet works with the materials of social relationships. Composition, translation, also 'revision,' hence offering maximum opportunity for the resources of *criticism*" (173; emphasis in the original). From this "comic" perspective, theory does not provide a ground for praxis by discovering foundational or normative principles, but rather provides "damage control" for praxis—damage control that is crucial because praxis is *always* of necessity "blind," always ungrounded, always reductive of difference or (in systems theory parlance) of an outside environment that is always already more complex than the system itself.

And at this precise juncture, the theoretical challenge then becomes how to acknowledge this without falling immediately back on an uncritical pluralism that says that all contexts are equal because they are all equally constraining in formal terms—a problem, as I have argued, that is especially acute for systems theory. As Jameson puts it in the conclusion to *Postmodernism:*

> The very concept of difference itself is booby-trapped....Much of what passes for a spirited defense of difference is, of course, simply liberal tolerance, a position whose offensive complacencies are well known but which has at least the merit of raising the embarrassing historical question of whether the tolerance of difference, as a social fact, is not the result of social homogenization and standardization and obliteration of genuine social difference in the first place. (341)

What this means is that we must avoid not only an uncritical dogmatism, but an uncritical pluralism as well. We must resist not only dogma, in other words, but also the dogma of no dogma at all; we must be open to what we might call "strategic totalization" if we are to disclose, for example, what dialectical thought calls the "contradiction" between equal abstract rights in the legal sphere and the real inequality under the law made possible by the asymmetrical distribution of *property* rights. That sort of procedure is hinted at, if not exactly endorsed, in Jameson's latest book, *The Seeds of Time*, which seems a bit more reserved about (if still committed to) the dialectic. "My own feeling," Jameson writes,

> has been that, rather than positing a situation in which we have to choose between these two categories (contradiction standing for the modernist option perhaps, while antinomy offers a more postmodern one), it might be worthwhile using them both concurrently and against one another, insofar as each is uniquely equipped to problematize the other in its most vital implications.... [T]hey stand as each other's bad conscience, and as a breath of suspicion that clings to the concept itself.... [T]hese pointed reciprocal doubts can do the mind no harm and may even do it some good. (4)

From the vantage of the "comic" frame, the role of theory is not—indeed, cannot be in any coherent sense[20]—to *ground* politics and praxis, but rather to provide something like "unending critique," with the two "sides" of the problematic we have been sketching thus far serving as each other's "bad conscience" in a ceaseless antagonism.

This experimental, skeptical attitude is an important if often neglected part of the Marxist tradition.[21] It is useful to remind ourselves as well of the continuity between modernity and postmodernity, insofar as modernity, as Smart argues, as an "attitude or ethos" denotes "a permanent critique of ourselves and our era," the "relevance of a form of analysis which is simultaneously critical, historical, and experimental," which aims to "problematise our relationship to the present, illuminate through historical analysis the limits to which we are subject, and thereby open up the possibility of transgression" (161). But it is equally useful, perhaps, to specify a *difference* in "attitude or ethos" between the modern and the postmodern exemplified in the relationship between theory and the project of permanent critique that Smart, via the late Foucault, associates with modernism; for it is not only that we must reject the procrustean view of politics associated with modernity that says that one can engage in resistance and critique only if one first makes a universal, foundational, or normative claim. Nor is it only that that project cannot proceed, as Best and Kellner suggest, by way of a renewed commitment to a dialectical critical theory.

What is needed here, in other words, is not *just* the Burkean "comic" attitude or simply a well-meaning commitment to open-mindedness and self-critique; expressing the desirability of those attributes *is not the same as having a rigorous and coherent theoretical account of that desirability's necessity*. As we have already seen in our discussion of Walter Benn Michaels and the liberal problematic, one cannot simply "take thought" (to use a phrase favored by Jameson) and thereby bootstrap oneself out of the complexities and challenges foregrounded by theoretical investigation—a position tantamount, as Tom Cohen suggests, to Rorty's self-serving and ethnocentric claim that "we have enough theory" already. To assume as much is to make Michaels's mistake, only in reverse; it is to assume not that there is no free choice, but rather that choice is all too free.[22]

The renovation of pragmatism by systems theory and related strains of poststructuralism is invaluable not only in this regard, but also in helping us to realize at one and the same time that theory cannot ground politics in the way that modernity imagines because—to borrow once again Gasché's formulation—the *conditions of possibility* identified by theory are "at the same time *conditions of im-*

possibility."[23] What Gasché calls the "infrastructures" of theory—he has in mind, of course, the Derridean notions of "trace," "différance," and so on—enable the possibility of the Foucauldian "permanent critique" associated with modernity precisely *because they disable* it (4) by failing to ground or secure it, a failure associated with the epistemological skepticism of *post*modernity. As Gasché explains—and here the similarities with systems theory are striking indeed—"the law articulated by an infrastructure applies to itself as well. It has an identity, that is, a minimal ideality that can be repeated only at the price of a relentless deferral of itself" (7). Like the law of the paradoxical identity of any constitutive distinction that Luhmann borrows from George Spencer Brown, "What these laws establish, indeed, is that any ideality, identity, or generality, hinges on a prior doubling, pointing away from (self), and referral to an Other—in other words, on a prior singularization" (7).

This does not mean, however, that the affirmation of difference pure and simple is enough, for, as Gasché explains in his discussion of Heidegger's concept of *Versammlung* (or "gathering"): "To reject all gathering because it can turn into self-identical individuality, totality, or System is to close the doors of reflection and philosophical interpretation. Is this not to abort what gathering still holds out for the future, to reveal a lack of respect for what is to come, for what has never yet been present?" (20). It is on the basis of this reorientation of theory toward its conditions of (im)possibility, then, that theory can make good on the critical imperatives described in Foucault's reassessment of Enlightenment, because it is on this basis that the relationship between theory and a future yet to come can be reoriented away from dialectical closure—away from, as Jacques Derrida famously put it in "Structure, Sign, and Play," *arche* and telos[24]—and toward what Luhmann characterizes as the operationalization of difference, which dialectic *says* it values but *can* value, it turns out, only in its endless deferral of difference.

In this light, systems theory may be seen, as Luhmann puts it, as something like the "reconstruction of deconstruction,"[25] insofar as it examines the *pragmatic* effects of reorienting theory away from dialectic and toward the difference of identity and nonidentity, and shows how the failure of identity-based schemes may be operationalized systemically as a kind of necessity and success, and not *only* by systems that are either language- or text-based. That is to say, not only by systems that are human and/or, as Luhmann argues, human*ist*.[26] As we suggested in our discussion of Bruno Latour and the "hybrid" networks of postmodernity that include all sorts of nonhuman agents and actors, this crucial feature would seem to imply the priority of systems theory over deconstruction for the "new social movements" such as ecology and animal rights. And the posthumanism of systems theory would also distinguish it from the theory of social antagonism we find in Žižek, which remains fatefully tied to the figure of the Human and the Oedipal (and, variously in his work, the Hegelian and the Kantian) problematic, *even if* it transvalues humanism's ethical valences (as in Žižek's critical reassessment of the Freudian/Kantian disavowal of "the Thing"). Žižek's theory of social antagonism, in other words, can theorize the antagonism *of* humans, *for* humans, only.

The priority of systems theory in the "reconstruction of deconstruction" is suggested in a second, different register by Pierre Bourdieu's passing criticism of deconstruction in *The Field of Cultural Production:* that "by claiming a radical break with the ambition of uncovering ahistorical and ontologically founded essences, this critique is likely to discourage the search for the foundation" of social forms and institutions where they are "truly located, namely, in the *history*" of those forms and institutions.[27] Luhmann has provided his own version of this historical emergence, of course, in his influential theory of "functionally differentiated" society. According to Luhmann, the transition to modern society is characterized above all by the movement away from stratified or hierarchical organization, in which the absolute monarch, the church, the court, or the aristocracy represents society as a whole, and toward a society of operationally closed, self-referential function systems — the legal system, the economic system, the education system, and so on. Under functional differentiation, as Schwanitz puts it, "society is no longer regarded as the sum of its parts, but as a combination of system-environment differentiations, each of which reconstructs the overall system as a unity for the respective subsystem and its specific environment according to the internal boundary of the subsystem" ("Systems Theory according to Niklas Luhmann" 144). Luhmann's account of functional differentiation gives us a picture of modern (and postmodern) society as a horizontal plane on which the different autopoietic function systems exist side by side, with no one system (the economic in the Marxist account, say) able to overdetermine the others. "The present state of world society," Luhmann writes, "can hardly be explained as a consequence of stratification. The dominant type of system-building within contemporary society relates to functions, not to social status, rank, and hierarchial order. The so-called 'class society' was already a consequence of functional differentiation, resulting, in particular, from the differentiation of the economic and the educational subsystems of society."[28]

For Luhmann, the attempt to define the unity of functionally differentiated society has historically resulted in a "trend toward increasing skepticism. The first idea," he writes, was, of course, that "division of labor" would increase welfare and produce a surplus available for new investment and/or for distribution.... The basic idea was now coherent modernization. If only society could succeed to modernize each of its function systems—to arrange for a market economy, for democracy, for universal literacy, for free "public opinion," and for research oriented by theory and method only (and not by social convenience)—then the hidden logic of functional differentiation (or invisible hand?) would grant success, i.e., an improved society.... [B]ut the preoccupation with these distinctions prevented the discussion of the question of why one could expect "modernized" function systems to support one another and to cooperate toward a better future. Nor did the neomarxist critique of modernization understand the problem, but rather turned back to a neohumanistic critique of class structures. But if system rationality depends upon a high degree of specialization and indifference, then how could one expect and even take for granted that "system integration" comes about? Would it not be more probable that developing systems would create difficulties, if not unsolvable problems, for each other—such as the internationalization of financial markets for any kind of socialist policy...? ("Why Does Society" 180–81)

For Luhmann, then, the "postmodern condition" constitutes not a break with the modern but rather an intensification of the systemic principles that organize the modern itself, resulting in our current predicament of "hypercomplexity": that is, "the availability in the system of a plurality of descriptions of the system" ("Why Does Society" 176)—the Luhmannian equivalent of Lyotard's "incredulity toward metanarratives"—and hence the increasing opaqueness of the system to itself generated by the incalculability of all the descriptive possibilities and their interactions.

  Luhmann's account of our current situation recalls much that we have already seen in Jameson's groundbreaking discussion of postmodernism—not least of all the collapse of "depth models" of knowledge and the loss of "critical distance" whereby one could assume a privileged perspective on difference and hypercomplexity. Viewed next to Jameson's model of the postmodern, however, Luhmann's reveals a fundamental weakness that I touched on earlier. Not least is the issue put on the front burner by Jameson's decision to place the economic system at the center of social organization and explanation. In the world of late capitalism, do we find persuasive or even plausible Luhmann's contention that the world we live in is one of horizontal functionally differentiated systems in which no system—most conspicuously in light of Jameson, of course, the economic—exerts a centrifugal force on the others? Are the social problems with which we are all familiar not related to the overdetermining fact of a dramatic redistribution of wealth upward from the working and middle classes to the wealthy since the late 1970s? How would Luhmann explain the relationship—which seems far from functionally differentiated and self-referential—between the economic and political systems that we see at work in political action committees, or the relationship between the economic, gender, and legal systems at work in the O. J. Simpson case? Luhmann might respond in these instances that such events reveal precisely the *imperative* toward functional differentiation—that is, the need to systemically insulate decisions of guilt or innocence from the amount of money available to the defendant. But such a reply would only reveal, as I have argued, the "liberal utopianism" that haunts Luhmann's valuation of complexity above all else. As William Rasch characterizes it:

> What presents itself, in Luhmann, as descriptive of modernity also takes on the force of a prescriptive. The description of modernity as differentiated needs to be read both as an empirical fact—"differentiation exists"—and as an imperative—"differentiation ought to (continue to) exist." That differentiation exists and ought to exist translates, then, into a political injunction: "Thou shalt not de-differentiate!"[29]

But the problem with Luhmann's account, of course, is precisely that it purports to be describing what *is* actually the case, not only what *ought* to be, and as such it imagines that in our society systems can engage in their own differential autopoiesis and the development of systemic complexity more freely than in fact they do.

As Jameson would no doubt be the first to point out, Luhmann's account reproduces all the problems of a liberal technocratic functionalism that has no way to address the sharp asymmetries of power in the social field, asymmetries that make the autopoiesis of social systems work better for some than for others. To say that social differentiation is perforce good is to immediately beg the question raised by Best and Kellner: that "some people and groups are in far better positions—politically, economically, and psychologically—to speak than others" (288). What is pressed by Jameson's analysis, problematic though it is, is once again the issue of structural causality, which wagers that some social systems—at least for the time being—exert more overdetermining force on the social field as a whole than others. For Jameson—though also not exclusively for Marxists—we do not in fact live in fully functionally differentiated society organized horizontally, but instead in a kind of hybrid society in which highly autonomous and self-referential forms usually associated with postmodernism (especially in areas like media and communications) coexist alongside more traditional hierarchical ones (such as the economic and class systems), which are associated with the stratified societies of early modernity, and which exercise an asymmetrical influence on the autopoiesis of other social systems. In such an account, the relation between the modern and the postmodern—between the twin Enlightenment legacies of administered society and permanent critique, Jameson's "scientific" totalization and Luhmann's liberal Utopian vision of full functional differentiation, Gasché's conditions of possibility and impossibility—might therefore be redescribed in terms (following Raymond Williams) of dominant, residual, and emergent historical trends.[30] On this view, as Best and Kellner put it, "we might want to speak of postmodern phenomena as only emergent tendencies within a still dominant modernity"—a modernity, to be sure, that follows Jameson's hierarchial account rather than Luhmann's differentiated one—"that is haunted as well by various forms of residual, traditional culture, or which intensify key dynamics of modernity, such as innovation and fragmentation" (279).

Such a hybrid account would present a more persuasive and compelling picture of our current situation than *either* Jameson's Marxist Utopian totalization in the name of the economic *or* Luhmann's liberal Utopian view of functional differentiation taken singly. And it would also provide a useful corrective to the post-Marxism of Ernesto Laclau and Chantal Mouffe, with whom I am otherwise, on theoretical and epistemological grounds, in large agreement. Laclau and Mouffe emphasize the contingency of all political identity and resistance, and insist that there is "no subject—nor further, any 'necessity'—which is absolutely radical and irrecuperable by the dominant order, and which constitutes an absolutely guaranteed point of departure for a total transformation" of society (*Hegemony* 169). Hence, their pragmatic insistence that "the hegemonic dimension of politics only expands as the open, non-sutured character of the social increases" (138), their effort to thoroughly break with the reductive monism, scientism, and economism of much of the Marxist tradition, their contention that social antagonism is irreducible—all of these are admirable and indeed crucial theoretico-political interventions at the current moment. Barry Smart is surely right, I think, to hold with Laclau and Mouffe that Marxism remains vulnerable in its conviction that "the social is sutured at some point, from which it is possible to fix the meaning of any event independent of any articulatory practice" (Laclau and Mouffe, *Hegemony* 177), a conviction whose corollary is "a privileging of particular antagonisms ('classism'), strategies ('economism'), and mechanisms ('statism')" (Smart, *Modern Conditions* 217).

What is more suspect in Laclau and Mouffe, however, is an overly sanguine view of the possibilities of resistance through liberalism and, indeed, consumerism—a problem only redoubled by their undertheorization of its relation to the socialism of their title in *Hegemony and Socialist Strategy*.[31] Laclau and Mouffe assume too readily a rapprochement between capitalism, liberalism, and the radical democracy they say they want, thereby failing to heed Jameson's warning about the overdetermining power of the economic system to direct and assimilate other forms of social difference—to systematically produce heteroglossia (to use Bakhtinian language) without dialogism. The problem lies not exactly with their contention that *"The task of the left therefore cannot be to renounce liberal-democratic ideology, but on the contrary, to deepen and expand it in the direction of a radical and plural democracy"* (Laclau and Mouffe 176). The problem is rather in their failure to recognize—as Jameson surely does—how this commitment to liberalism in the civil and social sphere is heavily compromised by the overdetermining power of *capitalism* in the economic. This is especially clear at revealing moments in *Hegemony and Socialist Strategy* such as the following:

> Interpellated as equals in their capacity as consumers, ever more numerous groups are impelled to reject the real inequalities which continue to exist. This "democratic consumer culture" has undoubtedly stimulated the emergence of new struggles which have played an important part in the rejection of old forms of subordination. . . . The phenomenon of the young is particularly interesting, and it is no cause for wonder that they should constitute a new axis for the emergence of antagonisms. In order to create new necessities, they are increasingly constructed as a specific category of consumer, which stimulates them to seek a financial autonomy that society is in no condition to give them. On the contrary, the economic crisis and unemployment make their situation difficult indeed. If we add to this the disintegration of the family cell and its growing reduction to pure functions of consumption, along with the absence of social forms of integration of these "new subjects" who have received the impact of the general questioning of existing hierarchies, we easily understand the different forms which the rebellion of the young has adopted in industrial societies.
>
> The fact that these "new antagonisms" are the expression of forms of resistance to the commodification, bureaucratization and increasing homogenization of social life itself explains why they should . . . crystallize into a demand for autonomy itself. (164)

It is possible, I suppose, to see this as an example of the unintended democratizing effects of consumer culture, but it seems much more plausible, given the analysis of the reification of difference under commodification available to us in the Marxist tradition, to see this as just one more example of how difference is "booby-trapped" (to use Jameson's phrase), of how consumer culture turns "the potentially revolutionary force of desire produced on capitalist terrain toward the work of conserving and perpetuating consumer capitalism."[32] What such an example seems to prove is rather the opposite of Laclau and Mouffe's point: how liberalism's desire for freedom and autonomy is channeled by consumer culture into the reproduction of an economic system that prevents the realization of those desires for a sizable majority of its members. In this light, Best and Kellner are right, I think, to observe of Laclau and Mouffe that "while democratic discourse may indeed have a 'subversive logic' that encourages people to demand their entitled rights and freedoms, they fail to analyze the ways in which capitalism can coopt or defuse these effects" (203)—a problem only exacerbated by the undertheorization of the relationship between democracy and socialism in Laclau and Mouffe's work (204).

Here again, the intervention of Deleuze and Foucault may prove of immense pragmatic value by fundamentally shifting the terms of the problematic away from the reification of difference by the system of commodification and toward the specific instances and bodily encounters within which the commodity system must perpetuate itself. Rather than focusing on how the interpellation of subjects as consumers might generate an expectation of equality, whose subsequent demystification and unmasking is somehow generalized as discontent across social space, Deleuze and Foucault focus our attention on the possibilities for resistance at work in the microdynamics of capitalist culture, which is uneven "all the way down," as it were, and not just in class terms. As Brian Massumi puts it:

> As powerful as the capitalist quasicause is, it remains a quasicause. It can only move into a prepared medium. It still relies on an army of despotic, disciplinary and liberal institutions to open bodies to it, to make them susceptible to its magic (armies, schools, churches, malls, . . .). These institutions concretize the capitalist relation. They determine that *this* purchase is made rather than another, that *this* activity or quantity of time is bought rather than another. They determine the particular forms of content taken by the capitalist relation, as superabstract form of expression.[33]

From the vantage of Deleuze and Foucault's analysis of the microdyamics of power and resistance, more spaces of difference and resistance are opened up to analysis, because the political dimension of social life is not limited to either democracy in civil society or its overdetermination by the putatively more fundamental level of class relations. That is to say, the *materialist* promise of Deleuze and Foucault's work is that it foregrounds the *outside* of any social practice or diagram—including the outside of capitalism that resonates in the concrete practices and spaces on which it is dependent for reproduction—as a reservoir of complexity and difference, a space of relatively "free" or "unbound" points, to use Deleuze's terminology. In that light, it would be overly generous, perhaps, but not altogether wrong, to say that Deleuze and Foucault's consumers may fare no better than Laclau and Mouffe's, but then, because the terrain of the problematic has been shifted and expanded, they do not need to.

In closing, I would like to turn briefly to a final set of questions in which the postmodern pragmatist account I have been developing might also be "comically" of use: not the question of the "hybrid" quality of postmodern society and its mixture of functional differentiation and structural causality by the economic, but rather the question of the vantage from which one can engage in a historicization of the emergent differences we find between a Jamesonian critique and a Luhmannian one. It is here, I think, that Jameson's famous dictum in *The Political Unconscious*—"always historicize!"—has, if rigorously pursued, unexpected consequences for his own account, and for the dialectical model on which it is based. This is so, as Jean-François Lyotard argues in some detail in *The Postmodern Condition*, because of changes in the conditions of knowledge themselves under postmodernism—changes that, as Smart correctly points out, the Marxist tradition has by and large ignored or treated with insufficient attention. As Smart observes, chief among these changes is what Anthony Giddens calls—in a formulation reminiscent of Luhmann's "hypercomplexity"—the "*reflexivity or circularity of social knowledge*" (quoted in Smart, *Modern Conditions* 193). In fields such as cognitive science, for example, it is not simply that changes in the social conditions of knowledge—in technologies, practices, and the very material factors of knowledge production in which Marxism should be interested—change *how* knowledge procedures are conducted; it is rather that those changes in turn transform *what* knowledge is and how we may interact with it and use it. An especially compelling case, as we have seen, is the field of cybernetics, where, in the years during and after World War II, our views of causality itself and of the relation between information and performance are radically altered—a redefinition of knowledge that is only intensified in the paradigms of complexity, self-organization, and emergence in second-wave cybernetics and, beyond that, chaos and complexity theory.

It is perfectly possible—indeed, it is entirely necessary—to make the historical materialist point that the paradigms of cybernetics arise out of the specific social, economic, and political conditions of the World War II effort, and more specifically in logistics and weapons research, that they are, in a sense, only possible in such a context (leaving aside the more daunting Marxist problem of whether these developments can be shown to be determined by the economic in the last instance). For example, one might very appropriately historicize the development of complexity theory, as Mitchell Waldrop and others have done, to show that the theorization of autocatalytic sets, so crucial to the early development of complexity theory as a whole, only becomes materially possible when a certain concentration and speed of computing power becomes available, so that mathematically simulated gene networks can economically and practically be put through enough computing cycles to determine whether or not complex but stable patterns of interaction emerge among them.[34]

But the larger point I wish to make is that *even if* we proceed along the lines of a more or less traditional form of materialist historicization, the changes we will identify in the social conditions and production of knowledge *do not leave the quality and character of that knowledge untouched*. In the case of complexity theory, for example, new paradigms of knowledge — ones with demonstrable pragmatic power and value, as the use of chaos theory in cardiology, meteorology, and economic analysis makes clear — do not simply question but *fundamentally undermine* the subject/object paradigm on which dialectics and the dialectical account of causality depend. How can we continue to believe in anything like a Marxist "science" when the very foundations on which that science bases itself have been radically questioned if not rendered obsolete by changes in the social conditions of knowledge and the new theoretical developments — like chaos theory, complexity theory, and systems theory — they have made possible? Nor are these changes limited to epistemological and philosophical consequences, for, as Anthony Giddens points out, what they mean is that

> no matter how well a system is designed and no matter how efficient its operators, the consequences of its introduction and functioning...cannot be wholly predicted.... New knowledge (concepts, theories, findings) does not simply render the social world more transparent, but alters its nature, spinning off in novel directions.... For all these reasons, we cannot seize "history" and bend it readily to our collective purposes.[35]

As Bruno Latour writes of the "ozone crisis" at the opening of *We Have Never Been Modern*, we

> discover that the meteorologists don't agree with the chemists; they're talking about cyclical fluctuations unrelated to human activity. So now the industrialists don't know what to do. The heads of state are also holding back. Should we wait? Is it already too late? Toward the bottom of the page, Third

World countries and ecologists add their grain of salt and talk about international treaties, moratoriums, the rights of future generations, and the right to development....

The horizons, the stakes, the time frames, the actors—none of these is commensurable, yet there they are, caught up in the same story.[36]

The problem with such a quintessentially postmodern predicament is that "the ozone hole is too social and too narrated to be truly natural; the strategy of industrial firms and heads of state is too full of chemical reactions to be reduced to power and interest; the discourse of the ecosphere is too real and too social to boil down to meaning effects."[37] In this light, the traditional Marxist concepts of mediation, homology, dialectic, and determination (even "in the last instance") by the economic seem woefully inadequate to the task at hand and, in an odd way, even reassuring. But it is a reassurance, as Latour suggests, that we must now forego, for if we are to take seriously the Jamesonian imperative to "always historicize!" then we must now do so with the addendum, "yes!—including the Marxist dialectic itself!"

Notes

Introduction. Nothing Fails like Success: The Postmodern Moment and the Problem of the "Outside"

1. Jean-François Lyotard, *The Postmodern Condition: A Report on Knowledge*, trans. Geoff Bennington and Brian Massumi (Minneapolis: University of Minnesota Press, 1984), xxiv. The postmodern turn in philosophy and criticism has been described in a whole host of studies. In the concluding chapter, I make explicit reference to two of the best: Barry Smart's *Modern Conditions, Postmodern Controversies* (London: Routledge, 1992), and Stephen Best and Douglas Kellner's *Postmodern Theory: Critical Interrogations* (New York: Guilford Press, 1991).

2. Jean-François Lyotard, *The Postmodern Explained: Correspondence 1982–1985*, trans. Don Berry et al., ed. Julian Pefanis and Morgan Thomas, Afterword Wlad Godzich (Minneapolis: University of Minnesota Press, 1993), 17–18, 29. I will not be discussing Lyotard's work in these pages in any detail, primarily because it is impossible to examine his confrontation with the "outside" of theory without engaging his concept of the sublime, which necessarily leads in turn to a protracted discussion of issues in aesthetics that would be out of key with my concerns here. At the same time, Lyotard's contention that science holds something like a privileged place in precipitating the crisis of postmodernism is very much to my purposes, as my discussion of systems theory will suggest. For Lyotard on science and postmodernism,

the reader should consult the whole of *The Postmodern Condition*. On the sublime, see, for example, what is probably Lyotard's single most important essay, "Answering the Question: What Is Postmodernism?" which is published as both an afterword to *The Postmodern Condition* and as the first chapter of *The Postmodern Explained*.

3. See William Rasch and Cary Wolfe, "Introduction: The Politics of Systems and Environments," *Cultural Critique* 30 (spring 1995): 5–13.

4. See Richard Rorty, *Philosophy and the Mirror of Nature* (Princeton, N.J.: Princeton University Press, 1979), 12.

5. Horace Fairlamb, *Critical Foundations: Postmodernity and the Question of Foundations* (New York: Cambridge University Press, 1994), 57.

6. Richard Rorty, *Objectivity, Relativism, and Truth*, Philosophical Papers, vol. 1 (New York: Cambridge University Press, 1991), 23–24. Further references are in the text. See also Barbara Herrnstein Smith, "Unloading the Self-Refutation Charge," *Common Knowledge* 2:2 (autumn 1993): 81–95, and, of course, the various documents associated with the ongoing Habermas/Luhmann debate, of which a helpful overview and reading are provided by Eva Knodt in "Toward a Non-Foundationalist Epistemology: The Habermas/Luhmann Controversy Revisited," *New German Critique* 61 (winter 1994): 77–100. As Knodt puts it, "if it can be shown that *any* attempt to ground a concept of rationality, whether one locates its ground in the constitutive powers of a transcendental subject or in a linguistically based notion of intersubjectivity, is fraught with as many logical difficulties [e.g., paradoxes and tautologies] as the critique of such projects, then the triumphant gesture of pointing out the paradox in an opponent's argument will lose its edge and reveal itself as mere rhetoric" (80).

7. See Cornel West, *The American Evasion of Philosophy: A Genealogy of Pragmatism* (Madison: University of Wisconsin Press, 1988).

8. Michael Taussig, *Mimesis and Alterity: A Particular History of the Senses* (New York: Routledge, 1993), xvii, xvi.

9. Brian Massumi, "The Autonomy of Affect," *Cultural Critique* 31 (winter 1995): 100.

10. William James, *Pragmatism* and *The Meaning of Truth* (one-volume edition) (Cambridge: Harvard University Press, 1978), xxiv.

11. Robert Lilienfeld, *The Rise of Systems Theory: An Ideological Analysis* (New York: John Wiley and Sons, 1978), 9, 10. Lilienfeld's distinction between the contextualist and the organic follows, as he notes, the typology of Stephen C. Pepper in *World Hypotheses—A Study in Evidence* (Berkeley: University of California Press, 1942; rpt. ed. 1970).

12. See James's examples of the squirrel going around the tree and of the cow path in *Pragmatism*, 27–28, 98. See, as well, Frank Lentricchia's discussion of the latter in *Ariel and the Police: Michel Foucault, William James, Wallace Stevens* (Madison: University of Wisconsin Press, 1988), 106–7.

13. Stanley Cavell, *This New Yet Unapproachable America* (Albuquerque, N.Mex.: Living Batch Press, 1989), 116.

14. See, for example, John McGowan, *Postmodernism and Its Critics* (Ithaca, N.Y.: Cornell University Press, 1991).

15. See Hilary Putnam, *Pragmatism: An Open Question* (Oxford: Basil Blackwell, 1995), 29–30, 38.

16. See Pierre Bourdieu, *Distinction: A Social Critique of the Judgment of Taste*, trans. Richard Nice (Cambridge: Harvard University Press, 1984), 494–98.

17. Stanley Cavell, "Postscript A: Skepticism and a Word concerning Deconstruction," in *In Quest of the Ordinary* (Chicago: University of Chicago Press, 1988), 131.

18. For a useful overview and history, see Steve Joshua Heims, *Constructing a Social Science for Postwar America: The Cybernetics Group 1946–1953* (Cambridge: MIT Press, 1993).

19. See, for example Donna Haraway's famous "Cyborg Manifesto," in *Simians, Cyborgs, and Women: The Reinvention of Nature* (New York: Routledge, 1991), 149–81.

20. Peter Galison, "The Ontology of the Enemy: Norbert Weiner and the Cybernetic Vision," *Critical Inquiry* 21:1 (autumn 1994): 251, 266.

21. Donna Haraway, "Situated Knowledges," in *Simians, Cyborgs, and Women*, 192.

22. Michael Hardt, *Gilles Deleuze: An Apprenticeship in Philosophy* (Minneapolis: University of Minnesota Press, 1993), 120–21.

23. John Rajchman, *Michel Foucault: The Freedom of Philosophy* (New York: Columbia University Press, 1985), 6–7.

24. Kenneth Burke, *Attitudes toward History*, 3d ed. (Berkeley: University of California Press, 1984), 173, 171; emphasis in the original.

25. Barbara Herrnstein Smith, "The Unquiet Judge: Activism without Objectivism in Law and Politics," in *Rethinking Objectivity*, ed. Alan Megill (Durham, N.C.: Duke University Press, 1994), 295.

1. Pragmatism: Rorty, Cavell, and Others

1. See William James, *Pragmatism* and *The Meaning of Truth* (one-volume edition) (Cambridge: Harvard University Press, 1978), 37, 126, and 301 for formulations close to Nietzsche's; also 123–24, 126, 290. On the contrast with "rationalism," 38, 124–25, 301–2. See also Frank Lentricchia, *Ariel and the Police: Michel Foucault, William James, Wallace Stevens* (Madison: University of Wisconsin Press, 1988), 110ff. Further references to both James and Lentricchia are given parenthetically in the text.

2. Walter Benn Michaels, "Saving the Text: Reference and Belief," *MLN* 93 (1978): 771. Further references are in the text.

3. Steven Knapp and Walter Benn Michaels, "Against Theory," in *Against Theory: Literary Studies and the New Pragmatism*, ed. W. J. T. Mitchell (Chicago: University of Chicago Press, 1985), 30. Further references are in the text. Jameson's description appears in *Postmodernism, or, The Cultural Logic of Late Capitalism* (Durham, N.C.: Duke University Press, 1990), 181–82. Further references are in the text.

4. Walter Benn Michaels, "Is There a Politics of Interpretation?" in *The Politics of Interpretation*, ed. W. J. T. Mitchell (Chicago: University of Chicago Press, 1983), 336. Further references are given in the text.

5. See also "Against Theory," 26–27, 29, which makes particularly clear the seamless and total character of "belief": "To imagine that we can see the beliefs we hold as no better than but 'merely different' from opposing beliefs held by others is to imagine a position from which we can see our beliefs without really believing them. To be in this position would be to see the truth about beliefs without actually having any—to know without believing" (27).

6. Walter Benn Michaels, *The Gold Standard and the Logic of Naturalism* (Berkeley: University of California Press, 1987), 177; my emphasis. Further references are in the text, abbreviated *GS*.

7. The most notable instance is perhaps to be found in the early work of Deleuze and Guattari. The object of Michaels's critique, however, is the position of Leo Bersani in *A Future for Astyanax* (1976): "desire," Bersani writes, "disintegrates society, the self, and the novel"; that is why, in Bersani's words, "the containment of desire is a triumph for social stability" (quoted in Michaels, *GS*, 47).

8. Evan Carton, "American Literary Histories as Social Practice," *Raritan* 8 (1989): 129–30. Further references are in the text.

9. Frank Lentricchia, *Criticism and Social Change* (Chicago: University of Chicago Press, 1983), 30. Further references are in the text, abbreviated *CSC*. Many contemporary Marxist critics would agree with Michaels's critique of desire here; they would differ, however about how that fact should be *interpreted*. As Jameson points out, Michaels's critique of Bersani would hold for a critic such as Julia Kristeva as well, and reaches its "paranoiac-critical" terminus in the work of Jean Baudrillard (*Postmodernism* 202–3).

10. The often-quoted passage in question on the relation of subject and interest in the concept of ideology is from Marx and Engels, *The German Ideology*, Part I: "For each new class which puts itself in the place of one ruling before it, is compelled, merely in order to carry through its aim, to represent its interest as the common interest of all the members of society, that is, expressed in ideal form: it has to give its ideas the form universality and represent them as the only rational, universally valid ones" (Karl Marx and Friedrich Engels, *The German Ideology*, ed. and intro. C. J. Arthur [New York: International Publishers, 1970], 65–66).

11. But for discussions of precisely that sort, see, among others, Terry Eagleton, *Ideology: An Introduction* (London: Verso, 1991); Göran Therborn, *The Power of Ideology and the Ideology of Power* (London: Verso, 1980); and, in a different register, *Ideology in Social Science*, ed. Robin Blackburn (New York: Pantheon Books, 1973). Excellent shorter introductions can be found in James H. Kavanagh, "Marxism's Althusser: Toward a Politics of Literary Theory," *diacritics* 12 (1982): 25–45, and in Myra Jehlen's introduction to *Ideology and Classic*

American Literature, ed. Sacvan Bercovitch and Myra Jehlen (Cambridge: Cambridge University Press, 1986), 1–18.

12. Martin Jay, "The Concept of Totality in Lukács and Adorno," *Telos* 32 (summer 1977): 130. For a more extended investigation of these matters, see Jay's study *Marxism and Totality: The Adventures of a Concept from Lukács to Habermas* (Berkeley: University of California Press, 1984).

13. For useful discussions of this point within Marxist theory, see Louis Althusser's canonical separation of the later from the early, "humanist" Marx in *For Marx*, trans. Ben Brewster (London: Verso, 1977). See also Perry Anderson, "Modernity and Revolution," *New Left Review* 144 (March–April 1984): 96–113, esp. 109–11, and Étienne Balibar, "The Vacillation of Ideology," in *Marxism and the Interpretation of Culture*, ed. and intro. Cary Nelson and Lawrence Grossberg (Urbana: University of Illinois Press, 1988), 159–209.

14. In fact, that sort of "interest-alignment" model of ideology which may be read in *The German Ideology* "is not Marxist," according to Althusser (and according, in his reading, to Marx's later work). See Louis Althusser, *Lenin and Philosophy and Other Essays*, trans. Ben Brewster (London: Verso, 1971), 127–86. This reconceptualization of the subject of ideology in light of Lacan's influence is already quite clear in Althusser's *For Marx* (in French, 1965). See especially the essay "Marxism and Humanism" in that volume.

15. Althusser, *For Marx*, 233–34.

16. The classic critique of the Lockean model of the subject is C. B. McPherson, *The Political Theory of Possessive Individualism: Hobbes to Locke* (Oxford: Oxford University Press, 1962). As for the theorization that stands behind McPherson's own, it is readily available in Marx's "On the Jewish Question," where he argues that the practical application of liberal "freedom" and "rights" is the right to private property. Hence, Marx tells us, the liberal self sees "in other men not the *realization* but the *limitation* of his own freedom" (Karl Marx, *Early Writings*, trans. Rodney Livingstone and Gregor Benton, intro. Lucio Colletti [New York: Vintage Books, 1975], 229–30).

17. This is not to suggest that this passage in *The Gold Standard* constitutes Michaels's only use of James. In fact, Michaels has often written about James, and seizes on other passages from *The Principles of Psychology* in *The Gold Standard* itself.

18. Quoted in Frank Lentricchia, "Philosophers of Modernism at Harvard, circa 1900," *South Atlantic Quarterly* 89:4 (fall 1990): 814; emphasis in the original. Further references are in the text.

19. The central example of this Burkean concept in action is to be found in Lentricchia's discussion of

Burke's proposed substitution of "the people" for "the worker" at the first American Writers' Congress in 1935 (*CSC* 21ff.).

20. This is, I take it, the point made by Michaels's former student Howard Horwitz, who chides historicist "formalism" for its "idolatry of forms as essence and prescription, the notion that an idea or symbolic form has only one use or means the same thing in all contexts." See Howard Horwitz, "The Standard Oil Trust as Emersonian Hero," *Raritan* 4:6 (spring 1987): 118.

21. Kenneth Burke, *Permanence and Change*, 3d ed. (Berkeley: University of California Press, 1984), 282. On the "bureaucratization of the imaginative," see Kenneth Burke, *Attitudes toward History*, 3d ed. (Berkeley: University of California Press, 1984), 225–26.

22. Frank Lentricchia, "On the Ideologies of Poetic Modernism, 1890–1913: The Example of William James," in *Reconstructing American Literary History*, ed. Sacvan Bercovitch (Cambridge: Harvard University Press, 1986), 245. This essay is an earlier version, different in some respects, of the treatment of James in *Ariel and the Police: Michel Foucault, William James, Wallace Stevens* (Madison: University of Wisconsin Press, 1988), and in the later "Philosophers of Modernism" essay. Further references to all of Lentricchia's studies are in the text.

23. Cornel West, *The American Evasion of Philosophy: A Genealogy of Pragmatism* (Madison: University of Wisconsin Press, 1989). Further references are in the text.

24. William James, *Essays in Pragmatism*, ed. Albury Castell (New York: Hafner, 1948), 144. Further references are in the text.

25. See Hilary Putnam's lecture, "The Permanence of William James," where he points out that "the view often attributed to James—that a statement is true if it will make people subjectively *happy* to believe it—is explicitly *rejected* by him." For James, truth, Putnam continues, "must be such that we can say how it is possible for us to grasp what it is," and hence for James "the notion of truth must not be represented as simply a mystery mental act by which we relate ourselves to a relation called 'correspondence' totally independent of the practices by which we *decide* what is and is not true." At the same time, however—and here is where he parts company with Rorty—Putnam writes that "unlike the pragmatists, I do not believe that truth can be *defined* in terms of verification" (Hilary Putnam, *Pragmatism: An Open Question* [Oxford: Basil Blackwell, 1995], 9, 10, 11). For Rorty's rejoinder and discussion of the differences between "verificationism" and Putnam's "technical realism," see the introduction to Rorty's *Consequences of Pragmatism: Essays 1972–1980* (Minneapolis: University of Minnesota Press, 1982), xxiii–xxix. Further references to Rorty's study are in the text.

26. For Putnam, "James' philosophy contains a strong strain of 'direct' realism, that is of the doctrine that perception is of objects and events 'out there,' and not of private 'sense data'" (*Pragmatism*, 19–20). Further references are in the text.

27. See, for instance, Giles Gunn, *Thinking across the American Grain: Ideology, Intellect, and the New Pragmatism* (Chicago: University of Chicago Press, 1992), who points out that pragmatism of the Rortyan variety "has come to be associated with cultural currents that are thought to be postliberal, if not antiliberal, in some very specific ways. It aligns itself...with the postmodernist and poststructuralist repudiation of culture as an expression of individual consciousness woven into patterns of consensus and dissent, of conformity and conflict, and it prefers to view culture as an intertextual system of signs that can be infinitely redescribed. It has thus positioned the critical recovery of pragmatist discourse essentially beyond the kinds of disputes that used to vex liberal criticism theoretically" (96). As will become clear, I view this fact from a vantage more or less opposed to Gunn's, which situates itself quite squarely within the terrain of liberal humanism. Further references to Gunn's study are in the text.

28. See Rorty's "Moral Identity and Private Autonomy: The Case of Foucault," in *Essays on Heidegger and Others*, Philosophical Papers, vol. 2 (Cambridge: Cambridge University Press, 1991), 193–98. As we shall see, Nietzsche's antifoundationalism must be for Rorty's liberalism cordoned off from what Rorty calls "the bad side of Nietzsche," the Nietzsche who does not exempt "Socratic conversation, Christian fellowship, and Enlightenment science" from his critique. See Richard Rorty, *Objectivity, Relativism, and Truth*, Philosophical Papers, vol. 1 (Cambridge: Cambridge University Press, 1991), 32. Further references to both of these works are given parenthetically in the text, abbreviated *EHO* and *ORT*, respectively.

29. See, for example, Slavoj Žižek's post-Lacanian analysis of the Look in a few different texts, most importantly *The Sublime Object of Ideology* (London: Verso, 1989), and, within feminism, the wealth of work by critics such as Mary Anne Doane, Laura Mulvey, Kaja Silverman, and many others.

30. Richard Rorty, *Philosophy and the Mirror of Nature* (Princeton, N.J.: Princeton University Press, 1979), 12. Further references are given in the text, abbreviated *PMN*.

31. Horace Fairlamb, *Critical Foundations: Postmodernity and the Question of Foundations* (New York: Cambridge University Press, 1994), 57.

32. Norman Geras, "Language, Truth and Justice," *New Left Review* 209 (January–February 1995): 110.

33. See also Barbara Herrnstein Smith, "Unloading the Self-Refutation Charge," *Common Knowledge* 2:2 (autumn 1993): 81–95, and, of course, the various documents associated with the ongoing Habermas/Luhmann debate, of which a helpful critical overview is provided by Eva Knodt in "Toward a Non-Foundationalist Epistemology: The Habermas/Luhmann Controversy Revisited," *New German Critique* 61 (winter 1994): 77–100. As Knodt puts it, "if it can be shown that *any* attempt to ground a concept of rationality, whether one locates its ground in the constitutive powers of a transcendental subject or in a linguistically based notion of intersubjectivity, is fraught with as many logical difficulties [e.g., paradoxes and tautologies] as the critique of such projects, then the triumphant gesture of pointing out the paradox in an opponent's argument will lose its edge and reveal itself as mere rhetoric" (80). See also, for Laclau's response to the charges leveled by Geras, Ernesto Laclau, *New Reflections on the Revolution of Our Time* (London: Verso, 1990), 103ff. Further references to Knodt's critique are given in the text.

34. See also *ORT* 202, where Rorty writes: "The view that every tradition is as rational or as moral as every other could be held only by a god, someone who had no need to use (but only to mention) the terms 'rational' or 'moral,' because she had no need to inquire or deliberate. Such a being would have escaped from history and conversation into contemplation and metanarrative. To accuse postmodernism of relativism is to try to put a metanarrative in the postmodernist's mouth."

35. In this connection, see also Rorty's Introduction to *Consequences of Pragmatism*, and the essay "The Contingency of Language" in *Contingency, Irony, and Solidarity* (Cambridge: Cambridge University Press, 1989), 3–22. Further references to this latter work are given in the text.

36. This is the point missed, it seems to me, in Norman Geras's reading of Rorty, which in general gives us a more traditionally realist extension of Putnam's reading. Geras rightly zeroes in on this moment in Rorty's argument, but mistakenly thinks that this constitutes Rorty's disavowal of his own position and consequent admission of realism. The faulty inference on Geras's part here is to assume that admitting that an "outside world" to which we are causally subject is the same as admitting that the realist account of the outside world was right all along. This assumption, as Malcolm Ashmore et al. have argued, "trades upon the objectivist assumption that rejecting realism is the same thing as rejecting everything that realists think is real." See Malcolm Ashmore, Derek Edwards, and Jonathan Potter, "The Bottom Line: The Rhetoric of Reality Demonstrations," *Configurations* 2:1 (winter 1994): 8.

37. See the essays by Fish and Knapp and Michaels in *Against Theory*, esp. 25, 113, 116. Although Rorty (whose response is included in the volume) agrees that theory (in the traditional sense) should "come to an end," as Knapp and Michaels put it, he differs on the following points: first, he does not agree with them that the separation of "meaning" and "intention" is an example of the theoretical enterprise; second, he finds that Knapp and Michaels, in their treatment of the Wordsworth poem, reintroduce the metaphysical distinction between real and accidental properties, between something that only "looks like" language and something that really is (see 102–8).

38. James, *Pragmatism* and *The Meaning of Truth*, xxiv; emphasis in the original. Further references are in given in the text.

39. See Richard Rorty, "Habermas and Lyotard on Postmodernity," in *Habermas and Modernity*, ed. Richard Bernstein (Cambridge: MIT Press, 1985), 161–75.

40. Nancy Fraser, *Unruly Practices: Power, Discourse and Gender in Contemporary Social Theory* (Minneapolis: University of Minnesota Press, 1989), 104. This is not to agree with Fraser's Habermasian call for a renewed attention to the normative. But Fraser's point is strikingly borne out in Rorty's response to Clifford Geertz's critique of his ethnocentrism, which we will discuss later in our examination of Michel Foucault. Further references to Fraser are in the text.

41. Chantal Mouffe, *The Return of the Political* (London: Verso, 1993), 10. It is an active question, of course, as to whether one can, historically or theoretically, affect the separation of political from economic liberalism in the way that Mouffe suggests—an issue to which I will turn in the final chapter.

42. Richard Rorty, "The Unpatriotic Academy," *New York Times* (February 13, 1994), section E, 15.

43. See Rorty, *ORT* 29–30 n. 11 on Deleuze, and *EHO* 193–94 on Foucault.

44. In this light, Rortyan pluralism would seem subject to the Deleuzian critique of "state philosophy" as described by Brian Massumi: "More insidious than its institution-based propagation is the State-form's ability to propagate *itself* without centrally directed inculcation (liberalism and good citizenship). Still more insidious is the process presiding over our present plight, in which the moral and philosophical foundations of national and personal identity have crumbled, making a mockery of the State-form—but the world keeps right on going *as if* they hadn't." See Brian Massumi, *A User's Guide to Capitalism and Schizophrenia: Deviations from Deleuze and Guattari* (Cambridge: MIT Press, 1992), 5. In this connection, see as well the discussion of the double sense of "representationalism"—as in "representative" democracy as well as in the presumed transparency of the sign to thing signified—in the conversation between

Foucault and Deleuze titled "Intellectuals and Power," in Michel Foucault, *Language, Counter-Memory, Practice* (Ithaca, N.Y.: Cornell University Press, 1977), 206.

45. Tom Cohen, "Too Legit to Quit: The Dubious Genealogies of Pragmatism," in *Anti-Mimesis from Plato to Hitchcock* (Cambridge: Cambridge University Press, 1994), 90. Further references are in the text.

46. Gunn's *Thinking across the American Grain*, for example, would be representative here.

47. Honi Fern Haber pursues something like this in her study *Beyond Postmodern Politics: Lyotard, Rorty, Foucault* (New York: Routledge, 1994). I applaud Haber's political commitments, her desire to come to terms with poststructuralism's theoretical challenge, and her impatience with the charge of relativism. Unfortunately, what Haber gives with one hand—the confrontation with poststructuralist antifoundationalism—she takes with the other, in her insistence on retaining the category of the subject or, even, the "self," in something very much like its traditional liberal humanist terms. For example, she writes that "I conclude with poststructuralism that human nature is always altered in creative ways" (115), when the real point of poststructuralism is that the idea of "human nature" is incoherent and pernicious. Haber wants to appropriate poststructuralism without having to abide by its posthumanist deconstructive rigor. As she puts it, "We must not allow the poststructural critique of language and the postmodern adoption of the law of difference to force us to conclude, as have some of its proponents, that there is no subject. In fact, my claim is that poststructuralism can be read—or *adapted* to read—as necessitating only the claim that there is no autonomous, wholly self-creating, or coherent *in the sense of one single-minded or one-track* self. The self can be many subjects" (120; emphasis in the original). As Cohen would be the first to point out, this division of self and subject undermines the materialist point of poststructuralism's commitment to exteriority, and thus reinstates the idea of liberal interiority (the "self" who pulls the strings of the various "subject" positions). Haber, in other words, would seem to fall midway on the spectrum between Gunn's liberal humanism and Cohen's "nihilistic" poststructuralism.

48. While the paradoxes of foundationalism are clear enough, it perhaps needs to be said that Rorty's constitutive paradox has to do with the crucial assertion that rescues the Rortyan pragmatist from idealism and makes way for theoretical reflection on an otherwise myopic belief. That paradox might be said to consist of Rorty's assertion that the pragmatist holds the all-constitutive belief that beliefs are not all-constitutive—a formulation for which he has been taken to task by Hilary Putnam in his *Pragmatism*, 74.

49. See Laclau, *New Reflections*, 104, 219–20.

50. Richard Fleming, "Continuing Cavell: Side Roads in *The Claim of Reason*," in Stanley Cavell et al., *Philosophical Passages: Wittgenstein, Emerson, Austin, Derrida* (Oxford: Basil Blackwell, 1995), 111. Further references to this work as a whole are given in the text, abbreviated *PP*.

51. One of the more notable discussions of these issues has centered around Ernesto Laclau and Chantal Mouffe's *Hegemony and Socialist Strategy* (London: Verso, 1985). For instructive discussions and/or critiques of that text, see Peter Osborne's essay "Radicalism without Limit?: Discourse, Democracy, and the Politics of Identity," in *Socialism and the Limits of Liberalism*, ed. Peter Osborne (London: Verso, 1991), 201–25. For a more detailed critique, see Ellen Meiksins Wood, *The Retreat from Class: A New "True" Socialism* (London: Verso, 1986).

52. Stanley Cavell, *In Quest of the Ordinary* (Chicago: University of Chicago Press, 1988), 131. Further references are in the text.

53. On the arrogation of voice, see chapter 1 of Cavell's *A Pitch of Philosophy* (Cambridge: Harvard University Press, 1995), esp. 3–10. The engagement with Derrida's reading of Austin takes place both in *Philosophical Passages*, in an essay and seminar on it titled "What Did Derrida Want of Austin?" (42–90), and in a more detailed version in the chapter "Counter-Philosophy and the Pawn of Voice," in *A Pitch of Philosophy* (53–128).

54. As Michael Fischer puts it in his book-length study of Cavell, all too often in deconstruction "epistemological assumptions keep freeing us from ethical dilemmas.... Because we cannot know others, we are relieved of the responsibility to read them accurately or face them. Because we are ineluctably hidden, we are not answerable for hiding." See his *Stanley Cavell and Literary Skepticism* (Chicago: University of Chicago Press, 1989), 77.

55. See also the Introduction to Stanley Cavell, *This New Yet Unapproachable America* (Albuquerque, N. Mex.: Living Batch Press, 1989), 23ff., and Cavell's essay "Politics as Opposed to What?" in *The Politics of Interpretation*, ed. W. J. T. Mitchell (Chicago: University of Chicago Press, 1983), 184ff. Further references to these works are given in the text.

56. Compare with Rorty's reading of Kant in "Nineteenth-Century Idealism and Twentieth-Century Textualism," in *Consequences of Pragmatism*, 139–59. That essay suggests, as does Rorty's later work even more so, that Rorty's answer to Cavell's question would be "yes."

57. This struggle with the philosophical terms of our existence is the subject of one of the more strange and remarkable passages in the whole of Cavell's work, one to which I cannot, unfortunately, do justice here. In *This New Yet Unapproachable America*, Cavell reads Emerson in "Experience" as offering the figure and fantasy of male childbirth as a kind of extended metaphor for the philosopher's attempt to invent, while writing it, the terms by which his own essay might be read — "an image of coming to terms as coming to term," as Cavell puts it. See, in particular, 92–93, 97, 102–3.

58. Stanley Cavell, "Thinking of Emerson," in *The Senses of Walden*, expanded ed. (San Francisco: North Point Press, 1981), 137. Further references are to this edition and are in the text.

59. Ralph Waldo Emerson, "The American Scholar," in *The Portable Emerson*, new ed., ed. Carl Bode with Malcolm Cowley (New York: Penguin, 1946), 63. Further references to Emerson's essays are from this edition and are given in the text.

60. I have argued this point in more detail in "Alone with America: Cavell, Emerson, and the Politics of Individualism," *New Literary History* 25:1 (winter 1994): 137–57. See esp. 142–44.

61. Emerson, "Self-Reliance," 142, 152.

62. Stanley Cavell, *Conditions Handsome and Unhandsome: The Constitution of Emersonian Perfectionism* (Chicago: University of Chicago Press, 1990), 198. Further references are given in the text.

63. Cavell's understanding of Utopian thought and its relation to presently existing society may be set in sharp and instructive contrast to Fredric Jameson's conception of the dialectic of utopia and ideology in *The Political Unconscious: Narrative as a Socially Symbolic Act* (Ithaca, N.Y.: Cornell University Press, 1981). We find the same sort of contradiction replayed at another key moment later in that same introduction. "In a democracy embodying good enough justice," Cavell writes, "the conversation over how good its justice is must take place and must also not have a victor . . . this is not because agreement can or should always be reached, but because disagreement, and separateness of position, is to be allowed its satisfactions, reached and expressed in particular ways. In the encounter of philosophy it is as important to keep still as to speak, to refuse sides, to wait" (*Conditions* 24–25). If we have this sort of responsiveness and willingness, then it is not clear why we need Emersonian perfectionism to teach us anything about democracy or justice *at all*, because to conduct this "conversation" at all we must have already attained those qualities that necessitate and are supposedly protected by that justice which was supposed to be the outcome of our discussion.

64. Emerson, *Nature*, in *The Portable Emerson*, 10.

65. Emerson, "Experience," in *The Portable Emerson*, 268. I have discussed the relationship between Emerson's individualism and the logic of property in more extensive detail in *The Limits of American Literary Ideology in Pound and Emerson* (Cambridge: Cambridge University Press, 1993). Further references to Emerson's essay are to this edition and are in the text.

66. Emerson, *Nature*, 27.

67. Sacvan Bercovitch, "Emerson, Individualism, and the Ambiguities of Dissent," *South Atlantic Quarterly* 89:3 (summer 1990): 645.

68. Michael Gilmore, *American Romanticism and the Marketplace* (Chicago: University of Chicago Press, 1985), 30, 31.

69. This is to disagree with Richard Grusin's reading of Emerson's relation to the rhetoric of property in his article " 'Put God in Your Debt': Emerson's Economy of Expenditure," *PMLA* 103 (1988): 35–44. Grusin is concerned to reject Marxist-influenced readings of Emerson's conceptual economy (such as that offered by Gilmore in *American Romanticism and the Marketplace*) in favor of more deconstructive interpretations informed by Baudrillard and Bataille. For Grusin, Emerson's "sacrificial economy of expenditure" is best read in terms of the logic of gift exchange or potlatch, not capitalist accounting (41); as Grusin puts it, "Virtue is not acquisition but expenditure; in sacrificing 'dead circumstances' [Emerson's phrase] one puts God in one's debt" (38).

The problem with this reading is that it is hard to see how we can speak of Emerson "sacrificing" circumstances when the "mid-world" of circumstance and experience is not held dear by Emerson in the first place, but is merely "*scoriae*" (*Nature*), "merest appearance" ("Self-Reliance"), or ephemeral and "counterfeit" ("Experience"). And even if this were not the case and Bataille's model of sacrifice were applicable to Emerson, the economy of potlatch or expenditure does not undo or deconstruct "acquisition," but only *defers* it. On this last point, see the critique of positions that reject the concept of utility offered by Barbara Herrnstein Smith in her study *Contingencies of Value* (Cambridge: Harvard University Press, 1989), chapter 6.

70. Perry Anderson, *In the Tracks of Historical Materialism* (London: Verso, 1984), 43–44.

71. See Jameson, *The Political Unconscious*, 10, 53–54, 193–94, 210–19, 266–68, 269–70.

72. McPherson, *The Political Theory of Possessive Individualism*, 142.

73. Karl Marx, "On the Jewish Question," in *Early Writings*, trans. Rodney Livingstone and Gregor Benton, intro. Lucio Colletti (New York: Vintage Books, 1975), 229–30.

2. Systems Theory: Maturana and Varela with Luhmann

1. Michel Foucault, "Truth and Power," in *The Foucault Reader*, ed. and intro. Paul Rabinow (New York: Pantheon Books, 1984), 58. Further references to this collection are in the text.

2. See, for example, Frank Lentricchia's chapter on Foucault in *Ariel and the Police: Michel Foucault, William James, Wallace Stevens* (Madison: University of Wisconsin Press, 1988).

3. As Slavoj Žižek notes, "Habermas and Foucault are two sides of the same coin," and "the Foucauldian notion of subject enters the humanist-elitist tradition" by way of the later Foucault's notion of the subject as "mastering the passion within himself and making out of his own life a work of art," "subject as the power of self-mediation and harmonizing the antagonistic forces, as a way of mastering the 'uses of pleasure' through a restoration of the image of self" (*The Sublime Object of Ideology* [London: Verso, 1989], 2). Further references are in the text.

4. Bruno Latour, *We Have Never Been Modern* (Cambridge: Harvard University Press, 1993). 136. Further references are in the text.

5. See in this connection two essays in *Socialism and the Limits of Liberalism*, ed. Peter Osborne (London: Verso, 1991): Ted Benton's "The Malthusian Challenge: Ecology, Natural Limits, and Human Emancipation" (241–69), and Kate Soper's "Greening Prometheus: Marxism and Ecology" (271–93).

6. See Theodor Adorno, *Negative Dialectics*, trans. E. B. Ashton (New York: Seabury Press, 1973), 183. See also Fredric Jameson, *Late Marxism: Adorno, or, The Persistence of the Dialectic* (London: Verso, 1990), esp. 20–21, 35–36, 96–99, 214–15; and Drucilla Cornell, *The Philosophy of the Limit* (New York: Routledge, 1992), esp. 16–24.

7. Michel Foucault, "Theatrum Philosophicum," in *Language, Counter-Memory, Practice*, trans. Donald F. Bouchard and Sherry Simon, ed. Donald F. Bouchard (Ithaca, N.Y.: Cornell University Press, 1977), 184–85. Further references are in the text.

8. In cognitive ethology and field ecology, see Donald Griffin, *Animal Minds* (Chicago: University of Chicago Press, 1992), and the essays collected in Marc Bekoff and Dale Jamieson, eds., *Interpretation and Explanation in the Study of Animal Behavior*, vol. 1 (Boulder, Colo.: Westview, 1990). In cognitive science and philosophy of mind, see Marian Stamp Dawkins, *Through Our Eyes Only?: The Search for Animal Consciousness* (Oxford: Freeman, 1993), and Daniel Dennett, *Consciousness Explained* (Boston: Little, Brown, 1991). And in animal rights philosophy, see Tom Regan, *The Case for Animal Rights* (Berkeley: University of California Press, 1983),

and Peter Singer, *Animal Liberation* (New York: Avon, 1975).

9. Donna J. Haraway, "A Cyborg Manifesto: Science, Technology, and Socialist-Feminism in the Late Twentieth Century," in *Simians, Cyborgs, and Women: The Reinvention of Nature* (New York: Routledge, 1991), 151–52. Further references are in the text.

10. Gayatri Spivak, "Remembering the Limits: Difference, Identity, and Practice," in Osborne, ed., *Socialism and the Limits of Liberalism*, 229. See also Étienne Balibar's essay "Racism and Nationalism," in which he observes that "in all these universals we can see the persistent presence of the same 'question': that of *the difference between humanity and animality.* . . . Man's animality, animality within and against man — hence the systematic 'bestialization' of individuals and racialized human groups — is thus the means specific to theoretical racism for conceptualizing human historicity" (in Étienne Balibar and Immanuel Wallerstein, *Race, Nation, Class: Ambiguous Identities*, trans. Chris Turner [London: Verso, 1991], 57).

11. Donna J. Haraway, "When Man™ is on the Menu," in *Incorporations*, ed. Jonathan Crary and Sanford Kwinter (New York: Zone Books, 1992), 43.

12. Donna J. Haraway, "Situated Knowledges: The Science Question in Feminism and the Privilege of Partial Perspective," in *Simians, Cyborgs, and Women*, 187. Further references are in the text.

13. The desire to hold on to the concept of objectivity is not by any means limited to feminist philosophy of science, of course. See, for example, George Levine's essay "Why Science Isn't Literature: The Importance of Differences," in *Rethinking Objectivity*, ed. Allan Megill (Durham, N.C.: Duke University Press, 1994), 65–79. See also Timothy Lenoir's discussion of a similar project in the work of Bruno Latour, in "Was the Last Turn the Right Turn? The Semiotic Turn and A. J. Greimas," *Configurations* 2:1 (winter 1994): 119–36.

14. Sandra Harding, "Introduction: Eurocentric Scientific Illiteracy — A Challenge for the World Community," in *The "Racial" Economy of Science*, ed. Sandra Harding (Bloomington: Indiana University Press, 1993), 17, 18. Further references are in the text.

15. Richard Rorty, *Objectivity, Relativism, and Truth*, Philosophical Papers, vol. 1 (Cambridge: Cambridge University Press, 1991), 6. Further references are in the text, abbreviated *ORT*. I borrow the distinction between "absolute" and "procedural" objectivity from Allan Megill's editorial introduction to *Rethinking Objectivity*.

16. Francisco Varela, Evan Thompson, and Eleanor Rosch, *The Embodied Mind: Cognitive Science and Human Understanding* (Cambridge: MIT Press, 1993), 202. Further references are in the text.

17. Barbara Herrnstein Smith, "The Unquiet Judge," in Megill, ed., *Rethinking Objectivity*, 295. Further references are in the text.

18. Malcolm Ashmore, Derek Edwards, and Jonathan Potter, "The Bottom Line: The Rhetoric of Reality Demonstrations," *Configurations* 2:1 (winter 1994): 11, 8. Further references are in the text.

19. Richard Rorty, "Habermas and Lyotard on Postmodernity," in *Habermas and Modernity*, ed. Richard Bernstein (Cambridge: MIT Press, 1985), 164. Further references are in the text.

20. In a way, this is simply to remind ourselves of the essentially ethical imperative of a certain brand of postmodern neo-Kantianism that insists, in thinkers as otherwise diverse as Lyotard and Habermas, that we respect the separation of discourses and the autonomy of language games. As is well known, that discursive difference and autonomy are subjected to a rather different fate in the end by Habermas and Lyotard, with the former insisting on the adjudication of knowledge claims by different discourses by the process of rational consensus, and the latter insisting that the intractable "differends" and "dissensus" between those different language games be respected, even at the price of abandoning any hope for consensus. For an overview, see Steven Best and Douglas Kellner, *Postmodern Theory: Critical Interrogations* (New York: Guilford Press, 1991). Further references to Best and Kellner are given in the text.

21. Evelyn Fox Keller, *Secrets of Life, Secrets of Death: Essays on Language, Gender and Science* (New York: Routledge, 1992), 74. Further references are given in the text.

22. Dietrich Schwanitz, "Systems Theory and the Environment of Theory," in *The Current in Criticism: Essays on the Present and Future of Literary Theory*, ed. Clayton Koelb and Virgil Lokke (West Lafayette, Ind.: Purdue University Press, 1987), 267. Further references are in the text.

23. Norbert Wiener, *Cybernetics, or, Control and Communication in the Animal and Machine*, 2d ed. (Cambridge: MIT Press, 1952), 5.

24. Ludwig von Bertalanffy, *General System Theory: Foundations, Development, Applications* (New York: George Braziller, 1968), 37. To take only two of the most well known examples of Bertalanffy's claim, it has long been widespread practice to use the explanatory model of the economic system to analyze the workings of ecosystems, where the explanatory mainspring is the investment, expenditure, and circulation not of capital but of *energy*. And the recent high-profile work in artificial intelligence and virtual reality takes for granted a comparative, cross-disciplinary deployment of systems theory crucial to early work in cybernetics: the use the same set of systemic

principles to compare the handling of information in binary computational systems with neuronal activity in the nervous systems of animals and humans. As Steve J. Heims explains the homology in his study of early cybernetics, early cyberneticians "made semi-quantitative comparisons between vacuum tubes and neurons, the overall size of brains and computers, their speed of operation and other characteristics": "Impulses arriving via axons from other neurons stimulate or in some instances inhibit a neuron from firing an impulse along its own axon. But the impulse, whenever it occurs, always has the same strength. Thus the firing of an impulse from a nerve cell can be conceived as a digital, binary process: A stimulus either generates an impulse or it does not. This fact is usually referred to as the all-or-none character of nervous activity. Like a piece of electronic equipment, the various characteristics of a neuron can be described quantitatively: A definite threshold voltage is required to stimulate a discharge; a certain "delay time" separates the arriving and the departing impulses; the impact of two arriving impulses will supplement each other provided they arrive within a well-defined, short time-span, the so-called period of latent addition; and so on." See Steve Joshua Heims, *Constructing a Social Science for Postwar America: The Cybernetics Group 1946–1953* (Cambridge: MIT Press, 1993), 20. Further references to both Bertalanffy and Heims are given in the text.

25. Robert Lilienfeld, *The Rise of Systems Theory: An Ideological Analysis* (New York: John Wiley and Sons, 1978), 11. Further references are given in the text. Lilienfeld's distinction between the contextualist and the organic follows, as he notes, the typology of Stephen C. Pepper in *World Hypotheses—A Study in Evidence* (Berkeley: University of California Press, 1942; rpt. ed. 1970).

26. Gregory Bateson, *Steps to an Ecology of Mind* (New York: Ballantine, 1972), 453. Further references are in the text.

27. See, for example, the work of economist Brian Arthur, who has worked extensively with the Santa Fe Institute on complexity theory. For a useful popular account, see M. Mitchell Waldrop, *Complexity: The Emerging Science at the Edge of Order and Chaos* (New York: Simon and Schuster, 1992).

28. Niklas Luhmann, "The Cognitive Program of Constructivism and a Reality That Remains Unknown," in *Selforganization: Portrait of a Scientific Revolution*, ed. Wolfgang Krohn et al. (Dordrecht: Kluwer, 1990), 72. Further references are in the text.

29. Heinz von Foerster, *Observing Systems*, 2d ed. (Seaside, Calif.: Intersystems, 1985), 258. Further references are in the text.

30. Gregory Bateson, *Mind and Nature: A Necessary Unity* (New York: Bantam, 1988), 8; emphasis in the original. Further references are in the text.

31. Ranulph Glanville and Franciso J. Varela, "Your Inside Is Out and Your Outside Is In (Beatles, [1968])," in *Applied Systems and Cybernetics. Proceedings of the International Congress on Applied Systems Research and Cybernetics*, vol. 2: *Systems Concepts, Models, and Methodology*, ed. G. E. Lasker (New York: Pergamon, 1980), 639. Further references are in the text.

32. On this point, see ibid., 639–40.

33. Félix Guattari, "The Three Ecologies," *New Formations* 8 (summer 1989): 141.

34. Franciso Varela, "The Reenchantment of the Concrete," in Crary and Kwinter, eds., *Incorporations*, 336. Further references are in the text.

35. Humberto Maturana and Francisco J. Varela, *The Tree of Knowledge: The Biological Roots of Human Understanding*, rev. ed., trans. Robert Paolucci (Boston: Shambhala, 1992), 242. Further references are in the text.

36. This view, in fact, is widely held in neurobiology (the scholarly field of research of Maturana and Varela) and in cognitive science, where philosophers such as Daniel Dennett agree with Maturana and Varela that "our world of colored objects is literally independent of the wavelength composition of the light coming from any scene we look at.... Rather, we must concentrate on understanding that the experience of a color corresponds to a specific pattern of states of activity in the nervous system which its structure determines" (*Tree* 21–22). See also Dennett, *Consciousness Explained*, esp. the chapter "Qaulia Disqualified" (which contains a section titled "Why Are There Colors?"). For further discussion of the example of vision by Maturana and Varela, see *Tree* 18–23, 126–27, and 161–62.

37. For a more detailed account, see Varela, "The Reenchantment of the Concrete," 332–35.

38. Quoted in Danilo Zolo, "Autopoiesis: Critique of a Postmodern Paradigm," *Telos* 86 (winter 1990–91): 67, 68. Further references are in the text.

39. Humberto Maturana, "Science and Daily Life: The Ontology of Scientific Explanations," in *Research and Reflexivity*, ed. Frederick Steier (London: Sage, 1991), 41–42; my emphasis. Further references are given in the text.

40. Niklas Luhmann, *Ecological Communication*, trans. John Bednarz (Chicago: University of Chicago Press, 1989), 23. Further references are in the text, abbreviated *EC*.

41. Niklas Luhmann, "The Autopoiesis of Social Systems," in *Essays on Self-Reference* (New York: Columbia University Press, 1990), 2. Further references are given in the text. See also the translator's introduction to Luhmann's *Ecological Communication*.

42. In Luhmann's work, this is part of the more general theorization, for which he is best known, of what he calls "functionally differentiated" modern society (as opposed to hierarchical or center/periphery premodern ones). For a rapid summary, see Luhmann's essay "The Self-Description of Society: Crisis Fashion and Sociological Theory," *International Journal of Comparative Sociology* 25:1–2 (1984): 59–72.

43. Luhmann in fact qualifies this somewhat in "Complexity and Meaning": "it has to be decided whether self-observation (or the capacity to handle distinctions and process information) is a prerequisite of autopoietic systems" (*Self-Reference* 82). It seems, though, that the position outlined earlier in the essay — that the concept of observation automatically includes that of self-observation — would require self-observation as such a prerequisite.

44. On the "liar's paradox," see Luhmann, *EC* xiv. Luhmann addresses the "theory of logical types" of Russell and Whitehead in many places; see, for example, "Tautology and Paradox" (*Self-Reference* 127) and, for a more extensive refutation, *EC* 23–24.

45. See also Luhmann, *EC* 37.

46. William Rasch, "Theories of Complexity, Complexities of Theory: Habermas, Luhmann, and the Study of Social Systems," *German Studies Review* 14 (1991): 70. As Rasch points out, this is precisely the point that is missed by Habermas's project of a universal pragmatics: "The whole movement of Habermas's thought tends to some final resting place, prescriptively in the form of consensus as the legitimate basis for social order, and methodologically in the form of a normative underlying simple structure which is said to dictate the proper shape of surface complexity" (78).

47. Of the lineage that runs from Nietzsche through Heidegger to Derrida, Luhmann writes that in their work "Paradoxicality is not avoided or evaded but, rather, openly exhibited and devotedly celebrated.... At present, it is not easy to form a judgment of this. Initially, one is impressed by the radicality with which the traditional European modes of thought are discarded.... [But it] has so far not produced significant results. The paradoxicalization of civilization has not led to the civilization of paradoxicality. One also starts to wonder whether it is appropriate to describe today's extremely dynamic society in terms of a semantic that amounts to a mixture of arbitrariness and paralysis" (Niklas Luhmann, "Sthenography," trans. Bernd Widdig, *Stanford Literature Review* 7[1990]: 134). Further references are given in the text.

48. As Luhmann points out, tautologies are actually "special cases of paradoxes"; "tautologies turn out to be paradoxes, while the reverse is not true" (*Self-Reference* 136).

49. As he puts it in "Cognitive Program," "The assumption—to be found above all in the classical sociology of knowledge—that latent structures, functions and interests lead to distortions of knowledge, if not to blatant errors, can and must be abandoned. The impossibility of distinguishing the distinction that one distinguishes with is an unavoidable precondition of cognition. The question of whether a given choice of distinction suits one's latent interests only arises on the level of second order observation [that is, on the level of the observation of observation]" (73).

50. We would want to note, for example, that Luhmann insists on the fundamental separation of psychic from social systems, a separation that Žižek would find simply absurd. On the other hand, Luhmann would object to Žižek's continued reliance on an essentially Enlighten-ment concept of the subject (even if its polarities of subject/object, mind/body are transvalued), which places the subject at the expressive center of the social system, insofar as all social relations are for Žižek essentially an exteriorization of a traumatic relation to the Real that is fundamentally a product of the internal psychic economy. On Luhmann and the separation of psychic and social systems, see Luhmann, "How Can the Mind Participate in Communication?" in *Materialities of Communication*, ed. Hans Ulrich Gumbrecht and K. Ludwig Pfeiffer (Stanford, Calif.: Stanford University Press, 1994), 371–87. On Žižek on the social as the exteriorization of an internal traumatic relation to the Real, see Slavoj Žižek, *Enjoy Your Symptom!: Jacques Lacan in Hollywood and Out* (New York: Routledge, 1992), 47–49.

51. Slavoj Žižek, "Beyond Discourse-Analysis," Afterword to Ernesto Laclau, *New Reflections of the Revolution of Our Time* (London: Verso, 1990), 249. Further references are in the text.

52. For a sketch of their differences, see Rorty's essay "Habermas and Lyotard on Postmodernity." As Rorty puts it, "the trouble with Habermas is not so much that he provides a metanarrative of emancipation as that he feels the need to legitimize, that he is not content to let the narratives which hold our culture together do their stuff. He is scratching where it does not itch" (164).

53. See, for example, the balanced account of Habermas's project in Best and Kellner's, *Postmodern Theory*, 233–55. See also Rasch's "Theories of Complexity," 70–72.

54. This is to leave aside Danilo Zolo's critique of the pragmatic value of Luhmann's work. As Zolo points out, it is so relentlessly abstract and circuitous that it is hard to imagine how anyone other than a quite sophisticated systems philosopher could make practical use of Luhmann's account. And even if we are suspicious of this charge—which might, it could be argued, harbor a latent anti-intellectualism—we must agree with Zolo that

Luhmann paints a picture of complexity so daunting that it is hard to imagine how one could ever, based on this theory, anticipate and direct decisions regarding the social system, as Luhmann himself freely, sometimes almost gleefully, acknowledges in many places. See Zolo, "Autopoiesis," 79.

55. Niklas Luhmann, "The Representation of Society within Society," in *Political Theory in the Welfare State*, trans. John Bednarz Jr. (Berlin: Walter de Gruyter, 1990), 17.

56. See, for example, the rather startling and willful naïveté of Luhmann's discussion of the relationship between politics and economics in ibid., 11–19; and for a typically reductive glance at Marxist theory, see 17–18 of that same essay.

57. See Jean-François Lyotard, "Dispatch concerning the Confusion of Reasons," in *The Postmodern Explained: Correspondence 1982–1985*, trans. Don Barry et al., ed. Julian Pefanis and Morgan Thomas, Afterword Wlad Godzich (Minneapolis: University of Minnesota Press, 1992), 66; see also 123–24. For a particularly striking instance of Rorty's technocratic functionalism, see his response to Clifford Geertz's critique of his ethnocentrism, in *ORT*, 203–10, which we will discuss in the following chapter.

58. John McGowan, *Postmodernism and Its Critics* (Ithaca, N.Y.: Cornell University Press, 1991), 198.

59. Richard Halpern, "The Lyric in the Field of Information: Autopoiesis and History in Donne's *Songs and Sonnets*," *Yale Journal of Criticism* 6:1 (1993): 208.

60. Donna J. Haraway, "The Biological Enterprise: Sex, Mind, and Profit from Human Engineering to Sociobiology," in *Simians, Cyborgs, and Women*, 44.

61. I appreciate Eva Knodt's observation along these lines in her response to an earlier draft of this chapter. See Raymond Williams's "Base and Superstructure in Marxist Theory," in *Problems in Materialism and Culture* (London: Verso, 1980). We will return to these issues in some detail in the concluding chapter.

62. See, for example, Haraway's "Sex, Mind, and Profit," Carolyn Merchant's *Radical Ecology* (New York: Routledge, 1993), Jeremy Rifkin, *Algeny* (New York: Penguin, 1984), and Peter Galison, "The Ontology of the Enemy: Norbert Wiener and the Cybernetic Vision," *Critical Inquiry* 21:1 (autumn 1994): 228–66. Further references to these studies are given in the text. Galison argues that cybernetics, in both Norbert Wiener and in Lyotard's postmodernism, represents "the apotheosis of behaviorism" (251), making "an angel of control and a devil of disorder" (266). In fairness, it should be noted that Haraway is ambivalent about the promise of the systems theory paradigm in a way that Galison is not. Although regretting that the cybernetic paradigm grew out of military research, she reminds us, with her usual good humor, that "illegitimate offspring are often exceedingly unfaithful to their origins. Their fathers, after all, are inessential" (150–51). While Galison insists that "the associations of cybernetics (and the cyborg) with weapons, oppositional tactics, and black-box conceptions of human nature do not so simply melt away" (260), a highly qualified claim that it is hard to imagine anyone disagreeing with, Haraway refuses to give up on the cybernetic paradigm precisely *because* of its imperative that we transgress the boundaries between the human and nonhuman, the organic and inorganic, the physical and nonphysical—boundaries that are in her view untenable, pernicious, and in any case nostalgic ways of thinking through our current dilemmas. It could be argued too that *first*-order cybernetics is a good deal more ambivalent and conflicted than Galison's essay suggests. Here, we might consult the fascinating "Introduction" to Norbert Wiener's *Cybernetics*, where he expresses great skepticism about the use of the cybernetic paradigm in the social sciences, and doubts "that sufficient progress can be registered in this direction to have an appreciable therapeutic effect on the present diseases of society." See Wiener, *Cybernetics*, 24.

63. See Fredric Jameson, *The Political Unconscious: Narrative as a Socially Symbolic Act* (Ithaca, N.Y.: Cornell University Press, 1983), 52–3, 59–60, 282–83. Further references are in the text.

64. Vincent Kenny and Philip Boxer, "Lacan and Maturana: Constructivist Origins for a Third-Order Cybernetics," *Communication and Cognition* 25:1 (1992): 95.

65. This theory of social antagonism would take issue not only with Maturana and Varela, but also with Jameson's positing—in *The Political Unconscious* and in essays like "Reification and Utopia in Mass Culture," *Social Text* 1 (winter 1979): 130–48—of a Utopian desire for collectivity as constitutive of the social and the projection of an "external enemy" (call him the capitalist) "who is preventing me from achieving identity with myself," when in reality this projection of an "other which is preventing me from achieving my full identity with myself is just an externalization of my own auto-negativity, of my self-hindering," which can never be abolished, "come the revolution" or otherwise (Žižek, "Beyond Discourse-Analysis" 252–53).

66. See in particular chapters 8 and 9 of *The Tree of Knowledge*, esp. 212, 224. For overviews of important recent work in cognitive ethology that reassesses these traits and behaviors in nonhuman animals, see Bekoff and Jamieson, Dawkins, Griffin, and the popular Jeffrey Moussaieff Masson and Susan McCarthy, *When Elephants Weep: The Emotional Lives of Animals* (New York: Delacorte, 1995). See also in particular *The Great Ape Project*, ed. Peter Singer and Paola Cavalieri (New York: St. Martin's Press, 1993).

3. Poststructuralism: Foucault with Deleuze

1. Michel Foucault, "Truth and Power," in *The Foucault Reader*, ed. Paul Rabinow (New York: Pantheon, 1984), 56. Further references are given in the text.

2. Brian Massumi, *A User's Guide to Capitalism and Schizophrenia: Deviations from Deleuze and Guattari* (Cambridge: MIT Press, 1992), 178 n. 74. Further references are in the text.

3. See Michel Foucault, *Remarks on Marx* (New York: Semiotext[e], 1991), 27, 31. See also James Bernauer, *Michel Foucault's Force of Flight: Toward an Ethics for Thought* (Atlantic Highlands, N.J.: Humanities Press, 1990). Further references to both of these works are in the text. It would be instructive as well to pursue the similarities between Foucault and Cavell, with an eye toward teasing out the stakes and motives in retaining (with Cavell) or rejecting (with Foucault) the image of the human in philosophy and theory.

4. Richard Rorty, *Essays on Heidegger and Others*, Philosophical Papers, vol. 2 (Cambridge: Cambridge University Press, 1991), 193. Further references are in the text, abbreviated *EHO*.

5. Richard Rorty, *Consequences of Pragmatism: Essays 1972–1980* (Minneapolis: University of Minnesota Press, 1982), 204, 205. Further references are given in the text.

6. Richard Rorty, "Foucault and Epistemology," in *Foucault: A Critical Reader*, ed. David Couzens Hoy (Oxford: Basil Blackwell, 1986), 47. Further references are in the text.

7. Richard Rorty, *Objectivity, Relativism, and Truth*, Philosophical Papers, vol. 1 (Cambridge: Cambridge University Press, 1991), 34. Further references are given in the text, abbreviated *ORT*.

8. Honi Fern Haber, *Beyond Postmodern Politics: Lyotard, Rorty, Foucault* (New York: Routledge, 1994), 90. Further references are in the text.

9. Quoted in Richard Rorty, *Contingency, Irony, and Solidarity* (Cambridge: Cambridge University Press, 1989), 63. Further references are in the text, abbreviated *CIS*.

10. Tom Cohen, *Anti-Mimesis from Plato to Hitchcock* (Cambridge: Cambridge University Press, 1994), 94. Further references are in the text.

11. See, for example, *ORT* 217, where Rorty lumps Marxism together with Christianity and Kantian Enlightenment as schemes that all traffic in teleological assurance in more or less the same way. The exceptions to this characterization of Marxism in Rorty usually occur when Rorty reads Marxism as a historicism congenial to pragmatist redescription (see, for example, *ORT* 198).

12. Michel Foucault, *Discipline and Punish: The Birth of the Prison*, trans. Alan Sheridan (New York: Vintage Books, 1979), 222. Further references are in the text.

13. Barry Smart, "The Politics of Truth and the Problem of Hegemony," in Hoy, ed., *Foucault: A Critical Reader*, 166. Further references are in the text.

14. For a critique of which see the opening section of Frank Lentricchia's *Criticism and Social Change* (Chicago: University of Chicago Press, 1983).

15. On the first point, see Rorty, *Consequences*, 208. On the avoidance of pain, see Rorty, *CIS*, 197.

16. It is entirely symptomatic, I think, that Rorty's essay in the collection *Mapping Ideology* zeroes in on the easy mark of the concept of ideology we find in Marx and Engels's *The German Ideology*, and leaves to the side the work on the materialist dimension of ideology in Marxism that is closest to that of Foucault himself. See Rorty, "Feminism, Ideology, and Deconstruction: A Pragmatist View," in *Mapping Ideology*, ed. Slavoj Žižek (London: Verso, 1994), 227–34.

17. Among such elements are Althusser's well-known insistence (in *For Marx*) that "Marxism is not a humanism," and his assertion (in the famous essay on ideological state apparatuses) that Marxist "science" (in contradistinction to ideology) is an essentially "subjectless" procedure. On Foucault's fitful relation to the Marxist tradition, see Abdul JanMohamed, "Refiguring Values, Power, Knowledge, or, Foucault's Disavowal of Marx," in *Whither Marxism?*, ed. and intro. Bernd Magnus and Stephen Cullenberg (New York: Routledge, 1995), 31–64.

18. See Foucault's "Truth and Power," 60–61, and *Discipline and Punish*, 135–38.

19. As is well known, Lukács places great emphasis on the proletariat's sudden achievement of critical class consciousness at moments of economic crisis, while Gramsci suggests that the worker's mind might achieve critical resistance and thus breed revolution even as his body is Taylorized and mechanized. On the distinction between materially oriented and phenomenologically oriented analyses of ideology, see Terry Eagleton's introductory chapter to his *Ideology: An Introduction* (London: Verso, 1994). On Gramsci and critical consciousness versus Foucault, see Frank Lentricchia, *Ariel and the Police: Michel Foucault, William James, Wallace Stevens* (Madison: University of Wisconsin Press, 1988), 70ff. And on Lukács, the proletariat, and crisis, see Edward Said, *The World, the Text, and the Critic* (Cambridge: Harvard University Press, 1983), 230–34.

20. See *Technologies of the Self: A Seminar with Michel Foucault*, ed. Luther H. Martin, Huck Gutman, and Patrick H. Hutton (Amherst: University of Massachusetts Press, 1988).

21. Louis Althusser, "Ideology and Ideological State Apparatuses (Notes Towards an Investigation)," in Žižek, ed., *Mapping Ideology*, 127.

22. Slavoj Žižek, "Introduction: The Spectre of Ideology," in *Mapping Ideology*, 13.

23. See Slavoj Žižek, *The Sublime Object of Ideology* (London: Verso, 1989), 30–35. Further references are given in the text.

24. It should be noted that there are important differences that I am ignoring in this triangulation of Žižek, Foucault, and Althusser, not the least of which, of course, is the fact that Žižek and Foucault have essentially opposed views of the relationship between psychoanalysis and ideology, with Althusser—one of whose crucial innovations is to link ideological "hailing" to the theory of the Lacanian Imaginary—as it were in the middle. Žižek addresses these differences in his introduction to *Mapping Ideology*, 13. As he puts it there, Foucault's attempt to bypass the notion of the problem of subject's investment in favor of a concept of power that inscribes itself directly on the body (and his consequent rejection of the term "ideology") contains "a fatal weakness" because it ignores, as Althusser does not, the psychoanalytic mechanisms needed to bridge "the abyss that separates micro-procedures from the spectre of Power"—"mechanisms which, in order to be operative, to 'seize' the individual, always-already presuppose the massive presence of the state, the transferential relationship of the individual toward state power, or—in Althusser's terms—towards the ideological big Other in whom the interpellation originates."

It should also be noted that Žižek has serious reservations about the move in Foucault's later work toward a recuperation of the notion of the Enlightenment subject and his focus on the *rapport à soi*. As he puts it in the introduction to *The Sublime Object of Ideology*, Foucault's claim in his later work that ethics consists of the fact that the subject must, without any foundational guarantees or universal rules, construct his own self-mastery, makes it clear, as Žižek puts it, that Foucault's notion of the subject is "a classical one: subject as the power of self-mediation and harmonizing the antagonistic forces, as a way of mastering the 'use of pleasures' through a restoration of the image of the self. Here," Žižek continues, "Habermas and Foucault are two sides of the same coin" (2).

This is not to suggest, however, that Žižek agrees wholesale with Althusser either. Žižek spells out his differences with Althusser in the section of *The Sublime Object of Ideology* titled "Kafka, Critic of Althusser," where he argues that the weakness of Althusser's theory is that it does not adequately think through the link between the process of ideological interpellation and the internalization of the Pascalian "machine." For Žižek, the answer to this problem is "that this external 'machine' of State Apparatuses exercises its force only in so far as it is experienced, in the unconscious economy of the subject, as traumatic, senseless injunction.... [W]e can learn from Pascal that this 'internalization,' by structural necessity, never fully succeeds, that there is always a residue, a leftover, a stain of traumatic irrationality and senselessness sticking to it, and that *this leftover, far from hindering the full submission of the subject to the ideological command, is the very condition of it;* it is precisely this non-integrated surplus of senseless traumatism which confers on the Law its unconditional authority" (43).

25. Jean-François Lyotard, *The Postmodern Explained: Correspondence 1982–1985*, trans. Don Berry et al., ed. Julian Pefanis and Morgan Thomas, Afterword Wlad Godzich (Minneapolis: University of Minnesota Press, 1993), 66. Further references are in the text.

26. Paul Rabinow, "Introduction," *The Foucault Reader*, 20. Further references are in the text.

27. "Polemics, Politics, and Problematizations: An Interview with Michel Foucault," in *The Foucault Reader*, 385. As Rorty responds to this passage in *Contingency, Irony, and Solidarity*, "That is, indeed, the problem. But I disagree with Foucault about whether in fact it is necessary to form a new 'we.' My principle disagreement with him is precisely over whether 'we liberals' is or is not good enough" (64).

28. See "Intellectuals and Power," in *Language, Counter-Memory, Practice*, trans. Donald F. Bouchard and Sherry Simon, ed. Donald F. Bouchard (Ithaca, N.Y.: Cornell University Press, 1977). Further references are in the text.

29. Michel Foucault, *Power/Knowledge: Selected Interviews and Other Writings, 1972–1977*, ed. Colin Gordon (New York: Random House, 1977), 82.

30. The late Foucault might well respond along the lines of an interview in which, when asked "What about someone who had sex so much he damaged his health?"—a sexual alcoholic, as it were—he answered "That's hubris, that's excess. The problem is not one of deviancy but of excess or moderation" (*Foucault Reader* 349). The problem with this response is that it opens him to the critique of his recovery of Enlightenment made by Žižek at the opening of *The Sublime Object of Ideology* (which we noted earlier).

31. Michel Foucault, *The Archaeology of Knowledge* and *The Discourse on Language*, trans. A. M. Sheridan Smith (New York: Pantheon Books, 1972), 17. Further references are in the text.

32. Bernauer responds to these charges by noting that "Foucault's actual work achieves something very different. Rather than promoting self-absorption, Foucault deprives the self of any illusion that it can become a sanctuary separated from the world.... Foucault's notion of self-formation is always presented in the context of a struggle for freedom within an historical situation," and so Foucault's self is not isolated but "agonistic" (as Foucault puts it); it becomes autonomous "only through a struggle with and a stylizing or adaptation of those concrete possibilities that present themselves as invitations for a practice of liberty" (181).

33. John Rajchman, *Michel Foucault: The Freedom of Philosophy* (New York: Columbia University Press, 1985), 6–7.

34. The connections between Foucault, James, and Burke are made implicitly and sometimes explicitly in the work of Frank Lentricchia. See especially his *Criticism and Social Change* (on Burke) and *Ariel and the Police* (on Emerson, Foucault, and James).

35. See Fredric Jameson's useful distinction between visions of "hard" and "soft" totality in modernity and postmodernity, in *The Political Unconscious: Narrative as a Socially Symbolic Act* (Ithaca, N.Y.: Cornell University Press, 1980), 92.

36. Gilles Deleuze and Claire Parnet, *Dialogues*, trans. Hugh Tomlinson and Barbara Habberjam (New York: Columbia University Press, 1987), vi. Further references are in the text.

37. On "state philosophy," see also Massumi, *User's Guide*, 4–5.

38. "Translators' Introduction" to Deleuze and Parnet, xii.

39. Gilles Deleuze, *Foucault*, trans. Sean Hand (Minneapolis: University of Minnesota Press, 1988), 122–23. Further references are given in the text.

40. Fredric Jameson, *Late Marxism: Adorno, or, The Persistence of the Dialectic* (London: New Left Books, 1990), 16. Further references are in the text.

41. Steven Best and Douglas Kellner, *Postmodern Theory: Critical Interrogations* (New York: Guilford Press, 1991), 107–8. Further references are in the text.

42. Fredric Jameson, "Marxism and Historicism," in *The Ideologies of Theory: Essays 1971–1986*, vol. 2, *Syntax of History* (Minneapolis: University of Minnesota Press, 1988), 161. Jameson's lengthiest reading of Deleuze and Guattari occurs in the essay "Beyond the Cave: Demystifying the Ideology of Modernism," in this same volume, esp. 123–28.

43. Fredric Jameson, "Pleasure: A Political Issue," in *The Ideologies of Theory*, vol. 2, 64.

44. Gilles Deleuze and Félix Guattari, *What Is Philosophy?*, trans. Hugh Tomlinson and Graham Burchell (New York: Columbia University Press, 1994), 7. Further references are in the text.

45. For more on the definition of "concepts," see ibid., 15–34.

46. Michael Hardt, *Gilles Deleuze: An Apprenticeship in Philosophy* (Minneapolis: University of Minnesota Press, 1993), 119. Further references are given in the text.

47. On the concept in Adorno, see Jameson, *Late Marxism*, esp. 15ff.

48. Todd May, "Difference and Unity in Gilles Deleuze," in *Gilles Deleuze and the Theater of Philosophy*, ed. Constantin V. Boundas and Dorothea Olkowski (New York: Routledge, 1994), 35–36. Further references are in the text.

49. As May puts it, Deleuze's "metaphysical claims are not claims about the way things are; rather, they are the structure of a new perspective. And his ethical claims—which are indeed ethical claims—are the articulation of a framework for thinking about other practices when one has taken up the perspective created by the concepts of a given metaphysics" (ibid., 39).

50. Hardt, *Gilles Deleuze*, 105. This sense of Deleuze's fundamental pragmatism is captured as well by Martin Joughin in his "Translator's Preface" to Deleuze's *Expressionism in Philosophy: Spinoza* (New York: Zone Books, 1990), where he writes that in Deleuze's books "the development of a 'philosophy' is traced from some version of an initial situation where some term in our experience diverges from its apparent relations which some other terms, breaking out of that 'space' of relations and provoking a reflection in which we consider reorientations or reinscriptions of this and other terms within a 'virtual' matrix of possible unfoldings of these terms and their relations in time.... Such a 'philosophy' comes full-circle when the 'subject'... 'orients' its own practical activity of interpretation, evaluation or orientation of the terms of experience within this universal matrix it has itself unfolded" (9).

51. As Deleuze points out by way of example, "this can clearly be seen even at a political party congress: the violence may be either in the hall or out in the street, while the ideology is always to be found on the platform; but the problem of the organization of power is settled privately in the adjoining room" (*Foucault*, 28).

52. Gilles Deleuze and Félix Guattari, *A Thousand Plateaus: Capitalism and Schizophrenia*, trans. Brian Massumi (Minneapolis: University of Minnesota Press, 1987), 66. Further references are in the text.

53. See also May, who contrasts the Deleuzian scheme with Derridean *différance:* "The latter involves the inevitable play of presence and absence, a specific economy of the two, which, although issuing in any number of philosophical possibilities, nevertheless governs them with a certain type of logic that is necessary to all discourse." Because of the complexities of Deleuzian "double articulation," however, for him there is no "guiding principle that underlies structures and that would thus be a unifying force determining them" (May, "Difference and Unity" 40).

54. As Hardt points out, "Deleuzian being is open to the intervention of political creations and social becomings: This openness is precisely the 'producibility' of being that Deleuze has appropriated from Scholastic thought. The power of society, to translate in Spinozan terms, corresponds to its power to be affected. The priority of the right or the good does not enter into this conception of openness. What is open, and what links the ontological to the political, is the expression of power: the free conflict and composition of the field of social forces" (*Gilles Deleuze* 120).

55. On "emergence" in Foucault's essay on Nietzsche, see *The Foucault Reader*, 82–86.

56. Brian Massumi, "The Autonomy of Affect," *Cultural Critique* 31 (fall 1995): 93. Further references are in the text.

57. As Hardt points out, the "virtuality" of the outside of thought—of what used to be called "nature"—constitutes something like Deleuze's fundamental ontological principle, "because the process of differentiation is the basic movement of life. *Elan vital* is presented in exactly these terms." As Deleuze puts it in *Bergsonism:* "It is always a case of a virtuality in the process of being actualized, a simplicity in the process of differentiating, a totality in the process of dividing: Proceeding 'by dissociation and division,' by 'dichotomy,' is the essence of life" (quoted in Hardt, *Gilles Deleuze* 15–16).

58. Michel Foucault, "Maurice Blanchot: The Thought from Outside," trans. Brian Massumi, in *Foucault/Blanchot* (New York: Zone Books, 1987), 55. Further references are in the text.

59. See also Hardt, *Gilles Deleuze*, 16–17.

60. N. Katherine Hayles, "Making the Cut: The Interplay of Narrative and System, or What Systems Theory Can't See," *Cultural Critique* 30 (spring 1995): 71.

61. Niklas Luhmann, "The Cognitive Program of Constructivism and a Reality That Remains Unknown," in *Selforganization: Portrait of a Scientific Revolution*, ed. Wolfgang Krohn et al. (Dordrecht: Kluwer, 1990), 76; my emphasis. Further references are in the text.

62. Constantin V. Boundas, "Deleuze: Serialization and Subject-Formation," in *Gilles Deleuze and the Theater of Philosophy*, 114. Further references are in the text.

63. Niklas Luhmann, "Deconstruction as Second-Order Observing," *New Literary History* 24 (1993): 769. Further references are in the text.

64. Niklas Luhmann, "The Paradoxy of Observing Systems," *Cultural Critique* 31 (fall 1995): 44. Further references are in the text.

65. See Gilles Deleuze, *Expressionism in Philosophy: Spinoza*, trans. Martin Joughin (New York: Zone Books, 1990), 13–22. Further references are in the text.

66. Gilles Deleuze, *Spinoza: Practical Philosophy*, trans. Robert Hurley (San Francisco: City Lights, 1988), 19. Further references are in the text.

67. See also ibid., 58–59.

68. Francisco J. Varela, "The Creative Circle: Sketches on the Natural History of Circularity," in *The Invented Reality*, ed. Paul Watzlawick (New York: Norton, 1984), 318; my emphasis. Further references are in the text.

69. See also in this connection Luhmann's use of the "folding" figure. For example, in "Deconstruction as Second-Order Observing," he describes something like the *reverse* of the folding process we have just examined: "Today logicians say that tautologies and paradoxes need *unfolding*—that is, they have to be replaced with stable identities. In one way or another one has to find distinctions that protect from the error of identifying what cannot be identified. But distinctions again become visible as paradoxes as soon as one tries to observe their unity. Unfoldments, then, are the result of unasking this question" (770). As Luhmann points out in *Ecological Communication* and elsewhere, both the theory of logical types of Russell and the binary coding used by systems are ways of not allowing the paradoxical identity of difference to be "identified." But both of these are themselves based on a constitutive paradox, which can only be disclosed by another observer. Hence, the only way that paradox can be unfolded is through the observation of observation.

70. Humberto Maturana and Francisco J. Varela, *The Tree of Knowledge: The Biological Roots of Human Understanding*, rev. ed., trans. Robert Paolucci (Boston: Shambhala, 1992), 169. Further references are in the text.

71. Ranulph Glanville and Francisco Varela, "Your Inside Is Out and Your Outside Is In (Beatles [1968])," in *Applied Systems and Cybernetics. Proceedings of the International Congress on Applied Systems Research and Cybernetics*," vol. 2, *Systems Concepts, Models, and Methodology*, ed. G. E. Lasker (New York: Pergamon, 1980): 640. Further references are in the text.

72. Humberto R. Maturana, "Science and Daily Life: The Ontology of Scientific Explanations," in *Research and Reflexivity*, ed. Frederick Steier (London: Sage, 1991), 34. Further references are in the text.

73. Alain Badiou, "Gilles Deleuze, *The Fold: Leibniz and the Baroque*," in *Gilles Deleuze and the Theater of Philosophy*, 52. Further references are in the text.

74. Gregory Bateson, *Steps to an Ecology of Mind* (New York: Ballantine, 1988), 460. Further references are in the text.

75. Gilles Deleuze, *The Fold: Leibniz and the Baroque*, trans. Tom Conley (Minneapolis: University of Minnesota Press, 1992), 81. Further references are given in the text.

76. Constantin Boundas and Dorothea Olkowski, "Editors' Introduction," in *Gilles Deleuze and the Theater of Philosophy*, 4. Further references are in the text.

77. For a pre-*Fold* overview of Deleuze's Leibniz, see Gilles Deleuze, *The Logic of Sense*, trans. Mark Lester with Charles Stivale, ed. Constantin V. Boundas (New York: Columbia University Press, 1990), 109–17.

Conclusion: Post-Marxism, Critical Politics, and the Environment of Theory

1. On these concepts in Louis Althusser, see his *For Marx*, trans. Ben Brewster (London: Verso, 1972), esp. the chapters "Marxism and Humanism," "Contradiction and Overdetermination," and "On the Materialist Dialectic." Further references are given in the text.

2. Abdul JanMohamed, "Refiguring Values, Power, Knowledge, or, Foucault's Disavowal of Marx," in *Whither Marxism?*, ed. Bernd Magnus and Stephen Cullenberg (New York: Routledge, 1995), 32.

3. On this point, see Steven Best and Douglas Kellner, *Postmodern Theory: Critical Interrogations* (New York: Guilford Press, 1991), 107. Further references are given in the text.

4. Barry Smart, *Modern Conditions, Postmodern Controversies* (London: Routledge, 1992), 218–19. Further references are in the text.

5. Fredric Jameson, *Postmodernism, or, the Cultural Logic of Late Capitalism* (Durham, N.C.: Duke University Press, 1991), 206. Further references are in the text.

6. Louis Althusser and Étienne Balibar, *Reading Capital*, trans. Ben Brewster (London: Verso, 1979), 300 n. 24. Further references are in the text.

7. See the opening chapter of Fredric Jameson, *The Seeds of Time* (New York: Columbia University Press, 1994). Further references are in the text.

8. "Fredric Jameson: Interview," *diacritics* 12 (1982): 80.

9. See Ernesto Laclau and Chantal Mouffe, *Hegemony and Socialist Strategy: Toward a Radical Democratic Politics* (London: Verso, 1985), 98–105. Further references are given in the text.

10. See the excellent overview of this line of development in Marxist theory from Althusser through the exhaustion of the concept of base and superstructure by Barry Hindess and Paul Hirst in James H. Kavanagh, "Marxism's Althusser: Towards a Politics of Literary Theory," *diacritics* 12 (1982): 25–45. See also Laclau and Mouffe, *Hegemony and Socialist Strategy*, 100–101.

11. For Jameson's influential discussion of this concept in *The Political Unconscious: Narrative as a Socially Symbolic Act* (Ithaca, N.Y.: Cornell University Press, 1981), see 24ff. Further references to Jameson's study are in the text.

12. On "expressive" versus "structural" concepts of totality and causality in Lukács, Althusser, and others, see Martin Jay's *Marxism and Totality: The Adventures of a Concept from Lukács to Habermas* (Berkeley: University of California Press, 1984) and, for an excellent shorter discussion, his "The Concept of Totality in Lukács and Adorno," *Telos* 32 (summer 1977): 117–37. Further references to Jay's book are in the text.

13. Fredric Jameson, *Late Marxism: Adorno, or, the Persistence of the Dialectic* (London: Verso, 1990), 6. Further references are in the text.

14. See Theodor Adorno, *Negative Dialectics*, trans. E. B. Ashton (New York: Seabury Press, 1973), 177–78. As Jameson would be the first to point out, we misunderstand Adorno's meaning of "nonidentity" if we do not recognize that Adorno often rearticulates the relationship of identity and nonidentity in terms of use value and exchange value, with the latter constituting something like the ur-form of identity generally. See Jameson's *Late Marxism*, 15–24.

15. See Kavanagh, "Marxism's Althusser," 27–28; as he makes clear, however, Althusser does not repudiate the earlier science/ideology distinction.

16. Louis Althusser, "Ideology and Ideological State Apparatuses," in *Lenin and Philosophy*, trans. Ben Brewster (New York: Monthly Review Press, 1971), 171. See also Sebastiano Timpanaro, *On Materialism*, trans. Lawrence Garner (London: Verso, 1975). Although the Althusserian notion of science and its internal verification procedures bears more than a passing resemblance to systems theory's insistence on the internal closure and self-reference of social systems, the signal difference is that, for Althusser, Marxist "science" provides a skyhook from which the relations of the different subsystems in the social totality may be seen as a whole—that, after all, is how Althusser can measure the extent of ideological "deformation" (as he puts it in *For Marx*) of "real" conditions of existence—whereas for a thinker like Luhmann no such authoritative, total view of society is possible.

17. Slavoj Žižek, "Beyond Discourse-Analysis," in Ernesto Laclau, *New Reflections on the Revolution of Our Time* (London: Verso, 1990), 259.

18. Michael Hardt, *Gilles Deleuze: An Apprenticeship in Philosophy* (Minneapolis: University of Minnesota Press, 1993), 120–21.

19. Kenneth Burke, *Attitudes toward History*, 3d ed. (Berkeley: University of California Press, 1984), 167. Further references are in the text.

20. For a perceptive overview of these issues, see Eva Knodt, "Toward a Non-Foundationalist Epistemology: The Habermas/Luhmann Controversy Revisited," *New German Critique* 61 (winter 1994): 77–100.

21. See, for example, Jameson's description of the twin concepts of "science" and "realism" in Bertolt Brecht: "The spirit of realism designates an active, curious, experimental, subversive—in a word, *scientific*—attitude towards social institutions and the material world; and the 'realistic' work of art is one which encourages and disseminates this attitude, yet not merely in a flat or mimetic way or along the lines of imitation alone" (Fredric Jameson, "Reflections in Conclusion," in *Aesthetics and Politics*, ed. Ronald Taylor [London: Verso, 1977], 205). See also Étienne Balibar's essay "The Vacillation of Ideology," which attempts to locate in the vacillation of the concept of ideology in Marx and Engels an attempt to do justice to the irreducible specificity and difference of contingent, pragmatic political practice that can never be reduced to the category of class interest as such. As Balibar puts it, "the theory, or rather the concept, of ideology denotes no other object than that of the nontotalizable (or nonrepresentable within a given order) complexity of the historical process," which implies in turn "the impossibility of a true fusion of theoretical and strategic functions" (in *Marxism and the Interpretation of Culture*, ed. Cary Nelson and Lawrence Grossberg [Urbana: University of Illinois Press, 1988], 203).

22. One would want, in other words, to add to Burke's "comic" frame two provisos that I have emphasized time and again in this study: that observing oneself while acting is both an imperative and an impossibility (because of the constitutive "blind spot" of observation), and hence that what is needed is a theory of the necessity of the observations of others, which systems theory attempts to provide; and second, that "maximum critical consciousness" on its own is not enough, as work on the materiality of ideology from Althusser through Foucault and Žižek (discussed in the preceding chapter) clearly demonstrates.

23. Rodolphe Gasché, *Inventions of Difference: On Jacques Derrida* (Cambridge: Harvard University Press, 1994), 4. Further references are in the text.

24. Jacques Derrida, "Structure, Sign, and Play in the Discourse of the Human Sciences," in *The Structuralist Controversy*, ed. Richard Macksey and Eugenio Donato (Baltimore: Johns Hopkins University Press, 1970), 247–65.

25. Niklas Luhmann, "Deconstruction as Second-Order Observing," *New Literary History* 24 (1993): 770. For an argument for the priority of systems theory over deconstruction, see Dietrich Schwanitz, "Systems Theory according to Niklas Luhmann — Its Environment and Conceptual Strategies," *Cultural Critique* 30 (spring 1995): 137–70, which contains a section titled "Systems Theory and Deconstruction." Further references to both essays are given in the text.

26. See Maturana and Varela's suggestion that the theorization of operation and observation in systems theory is in effect a more fundamental epistemological endeavor than that of deconstruction itself. As they put it in *Autopoiesis and Cognition:* "(5) The domain of discourse is a closed domain, and it is not possible to step outside of it through discourse. Because the domain of discourse is a closed domain it is possible to make the following ontological statement: *the logic of the* description *is the logic of the* describing *(living) system (and his cognitive domain).*
"(6) This logic demands a substratum for the occurrence of the discourse. We cannot talk about this substratum in absolute terms, however, because we would have to *describe* it, and a *description* is a set of interactions into which the *describer* and the listener can enter, and their discourse about these interactions will be another set of *descriptive* interactions that will remain in the same domain. Thus, although this substratum is required for epistemological reasons, nothing can be said about it other than what is meant in the ontological statement above" (Humberto R. Maturana and Francisco J. Varela, *Autopoiesis and Cognition: The Realization of the Living,* Boston Studies in the Philosophy of Science, vol. 42 [Dordrecht: D. Reidel Publishing Co., 1980], 39).

27. Pierre Bourdieu, *The Field of Cultural Production,* ed. and intro. Randal Johnson (New York: Columbia University Press, 1993), 255. As is well known, Bourdieu's famous critique of Derrida—which makes very much the same charge as the passage quoted from Luhmann—occurs in *Distinction: A Social Critique of the Judgment of Taste,* trans. Richard Nice (Cambridge: Harvard University Press, 1984), 494–98.

28. Niklas Luhmann, "Why Does Society Describe Itself as Postmodern?" *Cultural Critique* 30 (spring 1995): 180. Further references are in the text.

29. William Rasch, "Immanent Systems, Transcendental Temptations, and the Limits of Ethics," *Cultural Critique* 30 (spring 1995): 213.

30. A scheme that, as Best and Kellner point out, Jameson himself modifies and deploys, despite the impression his work gives of a sharp rupture between the modern and the postmodern (186). On the dominant, residual, and emergent, see Raymond Williams, *Marxism and Literature* (Oxford: Oxford University Press, 1977), 121–27.

31. It is not simply, as Best and Kellner have rightly pointed out (following earlier and more avowedly traditional Marxist critiques by Ellen Meiksins Wood and Norman Geras), that Laclau and Mouffe provide a rather misleading account of the Marxist tradition which suggests that it is a good deal more uniformly economistic and teleological than it is; nor is it only that they ignore critiques of these tendencies from within Marxism itself (200–201); nor is it simply that Laclau and Mouffe "reduce everything to discourse" and thereby "randomize" historical and social accounting—charges made by Wood and Geras that are equally reductive and misleading (202–3).

32. Frank Lentricchia, *Criticism and Social Change* (Chicago: University of Chicago Press, 1988), 30.

33. Brian Massumi, *A User's Guide to Capitalism and Schizophrenia: Deviations from Deleuze and Guattari* (Cambridge: MIT Press, 1992), 132–3.

34. See M. Mitchell Waldrop, *Complexity* (New York: Simon and Schuster, 1992), 106–12.

35. Anthony Giddens, *The Consequences of Modernity* (Stanford, Calif.: Stanford University Press, 1990), 153–54.

36. Bruno Latour, *We Have Never Been Modern* (Cambridge: Harvard University Press, 1993), 1.

37. Ibid., 6.

Index

Cary Wolfe is associate professor of English, American studies, and cultural studies
at Indiana University. Wolfe's many publications, including
The Limits of American Literary Ideology in Pound and Emerson (1993),
are devoted to critical theory and American literature and culture.